Stargazing
FOR
DUMMIES®

T0087690

by Steve Owens

A John Wiley and Sons, Ltd, Publication

Stargazing For Dummies®

Published by
John Wiley & Sons, Ltd.
The Atrium, Southern Gate,
Chichester
www.wiley.com

This edition first published 2013

© 2013 John Wiley & Sons, Ltd, Chichester, West Sussex.

Registered office

John Wiley & Sons Ltd, The Atrium, Southern Gate, Chichester, West Sussex, PO19 8SQ, United Kingdom

For details of our global editorial offices, for customer services and for information about how to apply for permission to reuse the copyright material in this book please see our website at www.wiley.com.

Wiley publishes in a variety of print and electronic formats and by print-on-demand. Some material included with standard print versions of this book may not be included in e-books or in print-on-demand. If this book refers to media such as a CD or DVD that is not included in the version you purchased, you may download this material at http://booksupport.wiley.com. For more information about Wiley products, visit www.wiley.com.

Designations used by companies to distinguish their products are often claimed as trademarks. All brand names and product names used in this book are trade names, service marks, trademarks or registered trademarks of their respective owners. The publisher is not associated with any product or vendor mentioned in this book.

For general information on our other products and services, please contact our Customer Care Department within the U.S. at 877-762-2974, outside the U.S. at (001) 317-572-3993, or fax 317-572-4002. For technical support, please visit www.wiley.com/techsupport.

A catalogue record for this book is available from the British Library.

ISBN 978-1-118-41156-8 (pbk); ISBN 978-1-118-41158-2 (ebk); ISBN 978-1-118-41160-5 (ebk); ISBN 978-1-118-41157-5 (ebk)

SKY10053987_082423

Contents at a Glance

Table of Contents

Introduction

●●●

Stargazing is a fascinating activity. For all of recorded history – and before that, no doubt! – people have looked up at the night sky and wondered what they were seeing.

For thousands of years stargazers had to make do with guesswork and make-believe, simply joining the dots and describing how the sky changed. But over the past four hundred years – ever since Galileo first turned a telescope to the night sky in 1609 – astronomers have begun to discover the countless wonders that fill the night sky, and to understand what they were looking at.

The sky changed from a canvas on which people drew pictures and told stories to a vast cosmos full of stars, planets, moons, galaxies, comets, asteroids, and beautiful clouds of dust and gas lit up by the stars around them. The universe contains so many incredible wonders that it has inspired generations of astronomers and stargazers to look upwards and wonder.

Stargazers – sometimes known as amateur astronomers – share an exciting hobby. Whether they stargaze on their own in their back gardens, or in clubs or societies, they explore what's overhead, becoming experts in the night sky.

There's always something interesting to look at too, from dramatic displays of northern lights or a total eclipse of the Sun, to the more everyday wonders of the rings of Saturn, the moons of Jupiter, or the beauty of a dark sky studded with thousands of stars.

About This Book

This book contains all you need to know about the fascinating hobby of stargazing. Using it you'll soon become an expert in identifying constellations, finding planets, hunting down faint elusive galaxies, and using the tools of the stargazer's trade – binoculars and telescopes. There's a lot in this book, with detailed descriptions of all of the 88 constellations visible from Earth (although you won't see them all at one time, or from one place) that'll help you become familiar with their patterns and where they appear throughout the year.

This book is intended as a reference guide, so you can dip into and out of it as you wish; you don't have to read it from cover to cover in the order laid out here. I explain what you need to know as you go along. Before long you'll be an expert stargazer, and the night sky will be filled with wonders that you'll be able to identify easily and point out to your friends, amazing them with your new-found knowledge and insight.

Conventions Used in This Book

As you use this book to explore the night sky, you'll find that I use the following conventions:

- ✔ The star charts in Part III of this book all have north uppermost, and depending on your stargazing location and what time of night you're observing them you might have to turn the book in your hands to make them match what you see in the sky.

- ✔ I have used *italic* text for the proper names for stars, but not for the planets or other celestial objects.

- ✔ `Monofont` text highlights a web address

What You're Not to Read

You'll occasionally find sidebars throughout this book. These shaded grey boxes have extra information in them that you might find interesting, but you can easily skip these without missing out anything essential. You might want to take more note of them in Chapters 10-15, where they give you handy information on the constellations you're looking for.

Foolish Assumptions

You're fascinated by the night sky, and have always dreamed of learning your way around the patterns of stars overhead. Maybe you can find a couple of familiar shapes – the Big Dipper, Orion – but that's about it; the rest of the sky is a confusing jumble of dots, and you want to begin exploring. Maybe you want to find out which planets are up when, or maybe you're looking to buy your first telescope to help you explore the sky more fully.

You're not a scientist, but you really want to understand your place in the universe, and see some of the wonderful objects that are dotted around the night sky. Whatever your goal, this book helps you achieve it.

How This Book Is Organised

You've probably already flicked through this book and noticed that it's divided up into four major parts.

Part 1: What's Up? Getting Familiar with the Night Sky

Every clear night you can see the stars overhead, and you're excited about stepping outside for your first proper stargazing expedition, but how do you make sense of what you see? How does the sky change over the course of the night, or from week to week, month to month, and over the year? What can you expect to see from your stargazing site, and how do you make sure you're safe and comfortable when outside at night peering skywards? And what equipment can you bring with you that will help you explore the sky in more detail?

This part gives you answers to all of these questions, helping you get the very most out of your new hobby, stargazing.

Part II: Joining the Dots: Learning Your Way Around the Night Sky

The huge variety of dots and fuzzy blobs up in the sky can be overwhelming at first. The chapters in this part explain what you're seeing when you look up at night, including stars and planets, the Sun and the Moon, the faint fuzzy patches of light that are galaxies and clouds of gas in space, the streaking lights of shooting stars, man-made lights of satellites, and much more besides.

You'll also find lots of useful information about how astronomers map the skies and divide it up into patterns of stars, the constellations.

Part III: Star Hopping

The bulk of the book is contained in this part, with detailed constellations guides for each of the 88 constellations that cover the entire sky. No matter where on Earth you are, or what season or time of night you're out stargazing, these guides will help you find everything of interest overhead, from the bright stars that make up the patterns of constellations, to the faint fuzzies that you'll need to chase down using your telescope.

This part is divided into six chapters, describing the constellations you find around the north pole and south pole regions of the sky, as well as the constellations visible in each of the four seasons.

Part IV: The Part of Tens

This part features two lists of things to look for in the night sky: your first ten astronomical objects, those that you will want to learn before you start exploring the rest of the sky, and the top ten dark sky objects, that you'll only see when you leave the bright city lights behind you.

Appendixes

This book has four online appendixes with useful information that will help you on your way as a stargazer. Appendix A has information for the position of the five naked-eye planets – Mercury, Venus, Mars, Jupiter, and Saturn – described month by month and year by year. Appendix B has a list of the constellations with their common names and other useful information. Appendix C lists the Messier objects, those hard-to-spot but great-to-find astronomical objects which range from nebulae to galaxies, and Appendix D gives you the info you need to enjoy the best of the meteor showers visible to the naked eye.

To find the Appendixes, head to www.dummies.com/go/ stargazingfd.

Icons Used in This Book

Throughout this book, some helpful icons appear to highlight useful information. Here's what each symbol means.

This icon is just a gentle reminder about something you've maybe read before, helping emphasise the point.

Signpost icons give you practical help on how to find your way around the night sky, using easy-to-spot stars and con-stellations to find the tougher ones.

Stargazing is a straightforward hobby, but an awful lot of com-plex science lies behind it. Skip this technical talk if you like, though you may find it more interesting than you'd imagine.

This target gives you the insider scoop that'll help you find exactly what you're looking for.

Stargazing is mostly about what you can *do*, but a few *do nots* crop up from time to time. This icon flags them up.

Where to Go from Here

You can jump into this book wherever you want. If you
want to buy a telescope then jump to Chapter 4, Your First
Telescope. Maybe you want to find out what constellations
are visible to you this season. In that case head straight to
Part III.

Wherever you begin I hope that you'll come to love the night
sky, and relish the opportunity to head out on a clear night
and explore the wonders of the universe over your head.

Part I
What's Up? Getting Familiar with the Night Sky

The 5th Wave
By Rich Tennant

"I've been thinkin' about starting a stargazing club, Dave. Wanna join?"

In this part...

The sky on a dark night is a fascinating sight, and that fascination has driven people throughout history to look up in wonder. They built monuments, like Stonehenge, to mark out the passage of time according to the skies. Even to this day the night sky still fills us with wonder.

In this part, I introduce you to the changing sky, how it changes over the course of a night, a week, a month, and a year, as well as guiding you through your first stargazing adventure, and the purchase of your first stargazing kit, such as binoculars or a telescope.

Chapter 1

The Changing Sky

As a stargazer, you'll spend a lot of time outside at night. Although the thousands of stars in the sky overhead may seem a little overwhelming at first, you'll quickly become used to the changing sky.

In this chapter, I look at the ways in which the sky changes, whether over the course of a night, several nights or throughout the year, and explore the reasons behind these changes.

Night and Day

Understanding the way the sky changes is an important part of stargazing. The earliest attempts to understand the universe were driven by the desire to mark out time, and ancient stargazers used the changing sky to do just that. Imagine that you're a stargazer thousands of years ago; what changes would you see happening in the sky, and how would you try to understand them?

The most obvious change in the sky happens every day as the Sun sets and the daylight fades, leaving a dark black sky studded with stars. If you have an accurate clock or stopwatch, you'll very quickly notice that the length of the day – that is, the daytime plus the night time – is the same every day. Add the time that the Sun spends above the horizon with the time it spends below the horizon, and you get one day of 24 hours.

Your spinning planet

Even the language used today harks back to the days before astronomers knew that the Earth was spinning and orbiting the Sun. The Sun appears to move across the sky, rising and setting, but it's actually the Earth spinning about its axis that gives people this mistaken impression. The Sun stays still; it's you that's moving.

Each day is 24 hours long because this is the time it takes for the Earth to spin once about its axis. If the Earth spun more slowly, your days would be longer; if it spun faster, your days would be shorter. (For example, on slow-spinning Venus, a day is 5,832 hours long, or 243 Earth days, while on fast-spinning Jupiter, the day is a shade under 10 hours long!)

Here comes the Sun (there goes the Sun)

Earth's local star, the Sun, gives you day and night as the Earth spins about its axis. The Earth spins anticlockwise if you're looking down on it from above the North Pole (see Figure 1-1), and so you see sunrise and sunset happening from specific directions.

The 24-hour day

Actually, the Earth doesn't spin once about its axis every 24 hours; it spins once every 23 hours, 56 minutes and 4 seconds. So where do the missing 3 minutes and 56 seconds go? They're not really missing at all. It's just that in the 23 hours, 56 minutes and 4 seconds that it takes for the Earth to spin, it's already moved a little around the Sun in its orbit.

Down here on Earth, a day is the length of time the Sun takes to make one complete circle of the sky – for example, the time it takes for the Sun to go from noon today to return to the exact same part of the sky at noon tomorrow. Between two consecutive noons as seen from Earth, the Earth has spun around on its axis once but then has to spin *a little bit farther around* for the Sun to get back to noon again. This 'little bit extra' spin takes 3 minutes and 56 seconds, and so the length of a day here on Earth – measured as the time between two noons – is 24 hours.

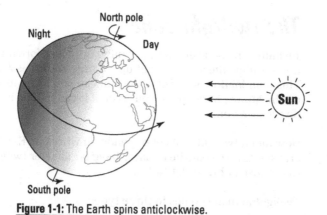

Figure 1-1: The Earth spins anticlockwise.

✔ **Sunrise:** Because the Earth spins anticlockwise, sunrise
  approaches from the east. Put another way, if the sky
  above your part of the Earth is just starting to brighten
  with the approach of dawn, somewhere a few hundred
  miles east of you, people will already have seen the Sun
  rising, while stargazers far west of you will still be enjoy-
  ing dark skies.

✔ **Midday:** Once the Sun rises above the horizon, it continues to climb in the sky until it gets to its highest point at
  midday, or noon. At this point, people in the northern
  hemisphere see the Sun due south, while those in the
  southern hemisphere see it due north.

✔ **Sunset:** After midday, the Sun begins to sink towards the
  west, gradually setting until it disappears below the horizon. Just as your friends in a city farther east than you
  see the Sun rise before you do, they'll also see it set first.

✔ **Midnight:** Once it's below the horizon, the Sun continues
  to light up the sky for a while before the sky darkens
  altogether. The Sun then keeps sinking lower and lower
  until it gets to its farthest below the horizon at midnight,
  at which point it begins making its climb back up to the
  next sunrise.

Stargazers near the equator see the Sun rising and setting at
nearly right angles to the horizon, while those farther north
or south of the equator see it rising and setting at a shallower
angle.

The twilight zone

Just after the Sun sets in the evening or just before it rises in the morning, the sky is still lit up – a time known as *twilight*. This illumination is because the Sun can still shine its light on the atmosphere above you, even if it can't shine its light directly on you.

How much twilight you get depends on where on Earth you are: stargazers near the equator have far shorter twilights than those at higher latitudes.

Twilight actually comes in three types:

- **Civil twilight:** Civil twilight is what most people mean when they talk about twilight. It starts in the morning when the Sun is six degrees below the horizon, and ends at sunrise. In the evening, civil twilight starts at sunset and continues until the Sun is six degrees below the horizon. During civil twilight, the sky is still bright enough that, in general, you don't need lights when doing things outside.

- **Nautical twilight:** Nautical twilight is when the Sun is between 6 and 12 degrees below the horizon. During nautical twilight, you can still distinguish the sky from the distant horizon when at sea, which allows sailors to take measurements of bright stars against the horizon (hence the name). Most people consider this time dark, but it's still technically twilight.

- **Astronomical twilight:** Astronomical twilight is when the Sun is between 12 and 18 degrees below the horizon. During astronomical twilight, you can no longer tell the sky from the distant horizon when at sea, but crucially for stargazers, light is still in the sky. Not much light appears – indeed, you'd probably say it's properly dark at this point – but the faintest objects in the sky, such as nebulae and very dim stars, will be visible only after astronomical twilight ends. Also, any long-exposure photographs of the night sky may show up this twilight that your eyes may miss.

Moonshine

After you've noticed the day and night cycle and the rising and setting of the Sun, the next most obvious change in the sky is the changing shape of the Moon. It's very obvious to anyone with even half an interest in what's going on overhead that the Moon looks different from night to night, and that sometimes it's fully round while at other times it shows a different shape, such as a crescent or a half Moon.

The phases of the Moon

Try this experiment: keep a record of the shape of the Moon each night. Draw the Moon as accurately as you can, but don't worry too much about detail. After you've been recording the Moon's shape for a few weeks, your drawing may look something like Figure 1-2.

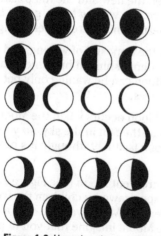

Figure 1-2: How the phases of the Moon look from Earth.

These changing shapes are the *phases of the Moon,* and they repeat in a regular pattern roughly every 29.5 days.

Each phase of the Moon has a name:

- New
- Waxing crescent

- First quarter (about a week after new)
- Waxing gibbous
- Full (about a week after first quarter)
- Waning gibbous
- Last quarter (about a week after full)
- Waning crescent
- New (about a week after last quarter)

The phase names look a little archaic, so you'll want to make sure you know what I mean when I say the Moon is waxing or waning, crescent or gibbous, new or full, and first or last quarter:

- **Waxing:** The adjective 'waxing' is used to describe a Moon that is growing in brightness – increasing in phase – from night to night. Consequently, the Moon can be waxing only between the new and the full phases. In the northern hemisphere, the waxing Moon grows from right to left; in the southern hemisphere it's left to right.

- **Waning:** Waning is the opposite of waxing, and describes the Moon when it's diminishing in brightness, or decreasing in phase. The Moon can be waning only between the full and the new phases.

- **New Moon:** A new Moon occurs when the Moon lies in the same direction as the Sun in the sky, so you can't see it at all. Half the Moon is still lit, though; it's just facing away from you.

- **Crescent Moon:** A crescent Moon (see Figure 1-3) is the familiar banana shape drawn by children the world over. When you see a crescent Moon, you see only a tiny sliver of the lit half of the Moon. Imagine that you're peeking around the edge, just catching a glimpse of the lit half.

- **Half Moon:** When the Moon is half way between new and full (and vice versa) it appears as a half Moon in the sky (see Figure 1-3). Just to confuse you, astronomers call these phases *first quarter* (when the Moon is between new and full) and *last quarter* (when it's between full and new). These names lead to all sorts of confusion. Why is it called a quarter Moon when it looks like half of a full Moon? The 'quarter' bit actually relates to how far the Moon is through its cycle of phases.

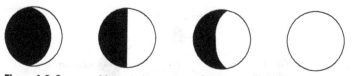

Figure 1-3: Crescent Moon, half Moon, gibbous Moon and full Moon.

- ✔ **Gibbous Moon:** The phase between quarter and full Moon – or between full and quarter Moon, depending on which part of the phase you're in – is called a gibbous Moon (see Figure 1-3). A gibbous Moon occurs when the Moon is a bit fatter than the quarter Moon but not quite full yet.

- ✔ **Full Moon:** The bright, round, full Moon (see Figure 1-3) is the midpoint of the cycle of phases; up until the full Moon, the Moon has been waxing, but after the full Moon, it begins waning back to new again, only to begin the cycle all over.

Your mantra: 'Half the Moon is always lit'

The Moon is a sphere of rock in space, and like the Earth, it's lit up by the Sun. As the Sun's light shines on the Moon, it lights up one entire hemisphere – half the Moon is always lit. But from Earth, you can't always see all of the lit half of the Moon, because the Moon orbits the Earth and changes position relative to the Sun (see Figure 1-4). In fact, you can only see all of the lit half when the Moon is full, when it's directly opposite the Sun in the sky. At all other times, you see only a fraction of the lit hemisphere of the Moon. The varying amount of the Moon's surface that you can see is what accounts for all the different phases. At all times, though, keep repeating to yourself: 'Half the Moon is always lit, half the Moon is always lit'!

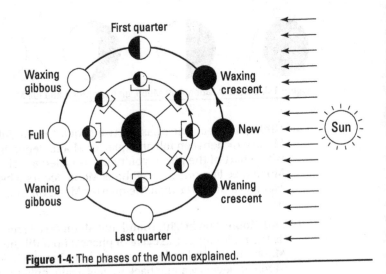

Figure 1-4: The phases of the Moon explained.

The 'moonth'

The calendar month is based on the phases of the Moon, which repeat every 29.5 days, which is roughly one month. In fact, the word 'month' derives from the word Moon. Technically, you should pronounce it 'moonth', but that just sounds silly.

Putting aside the fact that a month with half-days in it would be altogether confusing, the reason that months aren't exactly one *moonth* long (or to put it another way, the reason why the full Moon doesn't fall on the same day each month) is that for most people, the calendar has derived from the Greek and Roman calendar, which was a *solar calendar*, connected with the seasons and the year, rather than a *lunar calendar* based on the phases of the Moon. Having said that, the Hebrew and Islamic calendars (along with many others) are lunar, and so their feast days, for example, are often based on the phases of the Moon.

Once in a blue Moon

Because the Moon phases repeat every 29.5 days, and most months are 30 or 31 days long, you can have two full Moons in a calendar month; if one happens right at the start of a month, the second can sneak in right at the end. Some people call this a blue Moon, but the actual definition of a blue Moon is a bit more complicated than that description.

You can fit 12 lunar cycles of 29.5 days into a year of 365 days, and you'll be left with 11 days extra. As a result, every few years, you get 13 full Moons in a year rather than 12. This extra full Moon occurs as a fourth full Moon in a season of three, and this Moon is called the blue Moon.

The Changing Seasons

Ancient stargazers, once they'd noticed the changing day–night cycle and the Moon phases, also realised that the sky changes from season to season, too. The sky changes aren't as readily evident from one night to the next, but if you're out every night, the effects mount up.

The changes you'll notice if you stargaze regularly are:

✔ The Sun rises and sets at different times in different seasons: in winter, the Sun rises later and sets earlier than it does in summer.

✔ The length of day changes over the seasons (and the length of night, too).

✔ The Sun rises and sets at different points along the horizon in different seasons.

✔ These changes repeat every year (roughly every 365 days).

The Earth on tilt

All these changes happen because the Earth orbits the Sun once a year, and because the axis about which the Earth spins is tilted over (see Figure 1-5). In fact, the Earth's axis is tilted at 23.5 degrees to the straight up-and-down direction. This tilt gives you the seasons: if the Earth's axis tilted more, the seasonal changes would be more extreme; if it tilted less, they'd be less extreme. If the axis was perfectly perpendicular to the plane of orbit, then you'd have no seasons at all.

As the Earth orbits the Sun, and as it spins about its tilted axis, this axis always keeps pointing in the same direction. This means that in the northern hemisphere, the North Star, *Polaris,* which lies directly above the North Pole, stays in the same place in the sky.

As the Earth orbits the Sun, sometimes your hemisphere is tilted slightly towards the Sun, and at other times it's tilted away. The degree by which it's tilted towards or away from the Sun dictates your season. In midsummer, the tilt towards the Sun is at its greatest; in midwinter, the tilt away from the Sun is at its greatest.

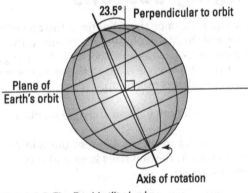

Figure 1-5: The Earth's tilted axis.

If you live in high latitudes, you'll notice these changes the most. Inside the Arctic and Antarctic circles, the seasonal changes are very extreme, with six months of daylight followed by six months of darkness. If you're at very high latitudes (but not quite inside the polar regions), you'll have long, dark winter nights followed six months later by long, bright summer days. If you're in mid latitudes, you'll notice less seasonal variation in length of day and season; if you're in the tropics or near the equator, you'll see virtually no difference at all in length of day or season.

One common misunderstanding is that it's hotter in summer than it is in winter because the Earth is nearer the Sun in summer. This belief isn't true; the heat is related to the tilt of the axis and whether your hemisphere is tipped towards or away from the Sun. The farther Earth's axis is tilted towards the Sun, the higher the Sun gets in the sky, and the hotter are your days (see Figure 1-6).

The Earth *does* sometimes get closer to the Sun, because its orbit isn't a perfect circle but rather an ellipse (a squashed circle). The date the Earth is closest to the Sun – a point called *perihelion* – is in early January, and the Earth is furthest from the Sun – at *aphelion* – in early July. This difference in distance from the Sun barely affects the temperature on Earth, though; any temperature difference is down to the height of the Sun in the sky and therefore the tilt of the axis.

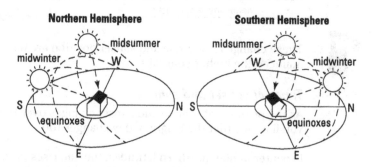

Figure 1-6: The Sun's height at midsummer and midwinter in the northern and southern hemispheres.

Sunset and sunrise, time and place

Because the degree to which your part of the Earth is tilted towards the Sun changes over the year, the Sun appears to rise and set at different points along the horizon.

Position of sunset and sunrise

Try this little experiment for yourself, especially if you have a clear eastern or western horizon: watch where the Sun rises or sets each day and remember that spot on the horizon.

Maybe the Sun rises behind a particular hill or tree on a certain day of the year. If you repeat this exercise every day, then over the course of months, you'll begin to see a pattern.

If you do this exercise for long enough, you'll notice that over the course of years, the sunrise and sunset positions swing back and forth along the horizon, with the Sun rising and setting at the same points on the same day each year (see Figure 1-7).

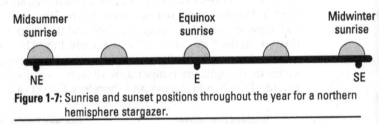

Midsummer sunrise   Equinox sunrise   Midwinter sunrise

NE   E   SE

Figure 1-7: Sunrise and sunset positions throughout the year for a northern hemisphere stargazer.

These changes in sunrise and sunset position are more pronounced the higher your latitude.

Times of sunset and sunrise

As the position of sunrise and sunset changes over the year, you'll notice that the length of day changes too.

In winter in high northern latitudes, the Sun rises in the southeast, barely lifting itself up above the southern horizon before setting again in the southwest In a southern winter in the southerly latitudes latitudes, the sun rises in the northeast and sets in the northwest. As a result, the days are short, and you get great long nights for stargazing.

In summer in high northern latitudes, the Sun rises in the northeast, climbing high above the southern horizon before setting again in the northwest. In the southern hemisphere the Sun rises in the southeast and sets in the southwest, but rises just as high. Consequently, you get long, bright days with very little darkness at night for stargazing.

This is why winter is often a stargazer's favourite season!

The Sun at a standstill

On two days each year, the Sun is said to be at *standstill,* on what astronomers call the *solstices.* These days usually fall on 20 or 21 June (northern summer, southern winter solstice) and 21 or 22 December (northern winter, southern summer solstice) each year.

The *summer solstice* occurs in midsummer. It's the day when the Sun climbs to its highest above the horizon, and when it spends longest above the horizon. For this reason, it's also known as the longest day.

The *winter solstice* happens in midwinter. It's the day when the Sun's noon position is the lowest it is all year, and when it spends the least time above the horizon. For this reason, it's also called 'the shortest day'.

Equinox

On two days each year, the Sun rises exactly due east and sets exactly due west, *everywhere on Earth,* and you get equal amounts of day and night. These two days are the *equinoxes.* These days usually fall on 20 or 21 March (northern spring, southern autumn equinox) and 22 or 23 September (northern autumn, southern spring equinox) each year.

The *spring equinox* (it's also called the *vernal equinox*) happens between the winter and summer solstices. The Sun spends 12 hours above and 12 hours below the horizon, before the days lengthen over the coming months as midsummer approaches.

Why 'standstill'?

The reason these solstices are called 'standstills' is that up until the summer solstice, the Sun has been gradually climbing higher in the sky at noon each day, spending a little bit more time above the horizon each day. Then, on the solstice, this gradual lengthening of the days stops – it's at a standstill. From then on, it's all downhill, with days shortening and the Sun rising lower and lower at noon each day, until the winter solstice when the Sun starts to climb again.

The *autumn equinox* happens between the summer and winter solstices. Like on the spring equinox, the Sun spends 12 hours above and 12 hours below the horizon, before the days shorten with the coming midwinter.

A very exact science

Although most people talk about solstices and equinoxes in terms of certain days, the reality is that they occur at an exact *instant* on a particular day.

A solstice is actually defined as the instant when the Earth's axis is most tilted towards or away from the Sun, while an equinox is defined as the instant when the Earth's axis is neither tilted towards or away from the Sun. This means that you'll often see the times of the solstices or equinoxes quoted to the nearest minute. However, it's still true to say that the day on which the solstice or equinox occurs is the summer solstice or vernal equinox, or whichever it happens to be.

The following table lists the solstices and equinoxes for 2012 to 2018.

Year	Solstices		Equinoxes	
2012	Jun 20 23:09	Dec 21 11:12	Mar 20 05:14	Sep 22 14:49
2013	Jun 21 05:04	Dec 21 17:11	Mar 20 11:02	Sep 22 20:44
2014	Jun 21 10:51	Dec 21 23:03	Mar 20 16:57	Sep 23 02:29
2015	Jun 21 16:38	Dec 22 04:48	Mar 20 22:45	Sep 23 08:21
2016	Jun 20 22:34	Dec 21 10:44	Mar 20 04:30	Sep 22 14:21
2017	Jun 21 04:24	Dec 21 16:28	Mar 20 10:29	Sep 22 20:02
2018	Jun 21 10:07	Dec 21 22:23	Mar 20 16:15	Sep 23 01:54

Chapter 2

Look Up! Your First Stargazing Trip

Maybe you've looked up at the stars on a clear night from where you live and wished you were brave enough to head somewhere darker, in a proper stargazing expedition. If so, this chapter is for you.

Unless you go well prepared and know what kind of stargazing spot to look for, chances are your stargazing trips will be cut short. You wouldn't go camping without checking your tent, and you shouldn't go stargazing without first making sure you have everything you need with you: warm clothes, a light to see by, a map of the area where you are and a map of the sky, too.

You also need to bear in mind a few important things when you venture on your first stargazing trip. You want to make sure you go somewhere where you'll get good views of the sky – that's what you're going out for, after all! But you can't just head out on any old night and expect to see amazing astronomical sights. You need to plan to have a good chance of seeing the stars – what's the local weather, what time does the Sun set and rise, and is the Moon up in the sky?

In this chapter, I assume that you aren't taking a telescope or a pair of binoculars with you and that you're just going to use your eyes. And for many stargazers, your eyes are all you need.

Getting Prepared to Catch a Glimpse of the Stars

You're about to head outside, maybe to some place remote, probably somewhere dark. Unless you're lucky enough to live somewhere with balmy nights, you're probably going out in the cold – maybe the very cold. So you need to prepare for this trip.

As with all things, safety comes first, so don't head anywhere that may be dangerous. If you're taking your car somewhere remote in the winter, then check the driving conditions and change your plans if necessary. The last thing you want is for your stargazing trip to turn into a bigger adventure than you meant it to be! Even if you're going out in a group – and especially if you're going out alone – let someone know where you're heading and when you expect to return.

Wrap up!

You don't want anything to spoil your magical view of the stars, especially nothing as mundane as the cold, so make sure you wrap up warmly. Standard stargazer attire consists of:

- ✔ Hat
- ✔ Gloves
- ✔ Scarf
- ✔ Thick socks
- ✔ Stout footwear
- ✔ Lots of layers
- ✔ Thermal underwear!

Seeing stars: Night vision

Your eyes are amazing, but they work better in bright light than at night, and they take a bit of time to adapt when going from daytime to nighttime.

Try this little experiment for yourself: wait inside until night-time and then walk from a brightly lit room into a dark room. How much can you see? Not a lot, I'm guessing. Now go back to the brightly lit room, turn the lights off and sit for ten minutes or so. Return to the second room. Can you see better now? You probably can, because of something called *dark adaptation* – the process your eyes go through in order to become better at seeing in the dark.

Very roughly speaking, two things happen in your eyes during dark adaptation:

1. **Your pupils – the black bits at the centre of your eyes that let in the light – get bigger to allow in more light.**

   That's good, because stars are pretty faint. In fact, the brightest star in the night sky – a star called *Sirius,* the Dog Star – is only as bright as a candle viewed from 200 miles away, so the more light your eyes can collect the better.

2. **Your eyes become more light-sensitive, thanks to a chemical called *rhodopsin*.**

   Usually you'll start to notice the effects of rhodopsin after a few minutes, but if you're out stargazing in the dark for longer, you'll *really* notice it: you'll be able to pick out stars and galaxies that you couldn't see at the beginning of the night.

Once you've got your night vision, you want to keep it! Any bright light shining in your eyes will mean you'll have to start dark adaptation all over again. This goes for the headlights of cars driving past, or someone turning on a bright light nearby. If you think that's about to happen, you should close your eyes until the light's gone.

The red light district

Say that you're outside gazing at the stars. You've got your night vision, having stood patiently for ten minutes waiting for the rhodopsin to do its thing, and then you drop your car keys. It's pitch black, and you can't find them. So turn on a torch, right? Wrong! At least don't turn on a bright white torch. Its glaring light will ruin your night vision. You need a red-light torch instead.

Red light is much less harmful to your night vision than white light, because your eyes aren't very reactive to red light. So you can turn on a red torch for a few minutes, find your keys, and then carry on stargazing.

You can find red-light torches for sale in specialist shops, but why buy one when you can make one? Get some red paper or cellophane and cut a piece to wrap around the front of your white-light torch and hold in place with a rubber band. You should use enough that the light coming out of your torch is very dim – only just bright enough for you to read by in the dark.

Knowing what to take

For your first naked-eye stargazing trip, you can travel light – a torch and warm clothes are all you really need – but you may also consider taking an insulated flask with a hot drink inside to help raise your spirits on a cold night.

A thick blanket to lay on the ground can come in handy, too, as lying down gives you a much better view of the whole sky and saves you a sore neck from all that craning skywards. Even more indulgently, consider a reclining lawn chair for truly luxurious stargazing! If you're lying down to stargaze, make sure you get up and move around regularly to stop yourself from getting too cold, though.

And, of course, you'll want to take a star map or guide so you know exactly where to look (see Chapter 9).

Identifying Your Local Stargazing Site

You can get a far better view of the night sky away from the bright lights of a town, but many people don't get the opportunity to travel to places with really dark skies for stargazing. Many times, you have to make do with what's on your doorstep. But don't be disheartened; by checking out your local neighbourhood for darker spots, you may be able to find somewhere that gives you a better view of the stars than you can get from the street where you live.

Most amateur astronomers have a favourite dark stargazing site, the place they head to for really good views. But on the nights when they can't drive for hours to set up their scopes, they'll head to their local stargazing site. Given that astronomers all look for the same things in a stargazing site, you may not be surprised to discover that other local astronomers use your spot, too. And if they don't, maybe you should spread the word. The more, the merrier when stargazing, and you can share tips about what you've learned.

The trouble with lights

Most people live in urban areas these days, and with the many conveniences of city living comes one major disadvantage: light at night. Whether it's from street lamps, car headlights, a nearby sports field, or your neighbour's security light, any bright light nearby produces *light pollution*.

Join your local astronomy society

There are plenty of great astronomy societies all around the world. If you want to learn the ropes, you should consider joining your local group. A quick Internet search should find you the society nearest you.

These groups tend to put on regular talks and stargazing evenings, so pop along to a few to find out whether you can see yourself becoming a member. Membership fees are usually very small – far less than the cost of a new telescope – and so astronomy societies provide an excellent and cost-effective way to get into stargazing as a hobby.

Light pollution comes in three forms:

✔ **Sky glow** is the combined effect of all the local night lights, and it does exactly what its name suggests – it makes the sky glow. In most cities, this glow is an orange colour. On a cloudless night, the sky overhead is lit up – and it isn't black, it's orange. The bigger the city you live in, the more lights surround you, and the brighter your sky will glow. There's not a lot you can do about sky glow short of leaving the city behind.

✔ **Glare** comes from nearby bright dazzling lights and instantly ruins your dark adaptation. You should avoid stargazing near any lights that may cause glare, even if they're not on when you first arrive. Maybe a bright security light will be triggered by a passing cat, or maybe the street lights near you are set to come on at a certain time of night.

✔ **Light trespass** is light that shines somewhere it isn't wanted. Light trespass may not be as glare-y as a bright light shining right at you, but any light that spills into your stargazing site can harm your night vision.

Usually problematic lighting is that which has been installed incorrectly (pointing up and out rather than down at the ground) or is too bright for the use it's intended for. If a nearby light is causing you problems, you can try speaking to the owner of the light. Perhaps the owner doesn't need it on all night or didn't realise it was causing problems. In the end, though, you may have to change your stargazing site if the lights around you are no good.

What you can expect to see

Sky glow from street lights sets a limit on what you can expect to see from your stargazing site. If you're stargazing in the middle of a big city, then the sky glow will be very bright indeed, drowning out the light from most stars. You'll still be able to see the Moon, the brighter planets, such as Venus, Mars, Jupiter and Saturn, as well as a few hundred stars that mark the more famous constellation shapes.

Campaigning for dark skies

After you get serious about stargazing, you'll begin to notice the profusion of bad lights that spoil your view of the night sky. Don't remain silent about them! You can contact your local council or city planners, write to your elected representative, and mention the problem to your neighbours. After all, it's not just stargazers who are affected by bright lights at night:

✔ It costs money to light up the sky unnecessarily.

✔ The energy used to produce that wasteful light probably came from burning fossil fuels that pollute the environment and contribute towards global warming.

✔ Light at night is harmful to nocturnal wildlife which relies on the natural day and night cycle for navigation and feeding.

✔ Evidence is beginning to arise that bright lights in the environment at night are bad for your health, too, and may even be carcinogenic.

You may want to join one of the many groups which are advocating better (not brighter, but better directed) light at night, such as the International Dark-sky Association and, in the UK, the Campaign for Dark Skies.

Stargazers in the suburbs of a bright city will see more stars than city dwellers will; you'll catch a glimpse of the Milky Way and some of the faint fuzzies that make stargazing so fascinating. You'll also see more satellites and meteors.

If you're ever lucky enough to observe from a rural site, then the splendour of the dark night sky will become apparent. You may still have to deal with some sky glow, but it will probably be limited to *light domes* on the horizon – halos of light that sit above large towns or cities off in the distance. In rural skies, you'll see a couple of thousand stars – the night sky can be so full of stars that it may take you longer to identify the constellations that you're learning – as well as the Milky Way stretching overhead in an arc and many faint fuzzy objects.

You can find truly dark skies in only a few remote places on the planet. You'll probably have to travel a long way to find dark skies, but the effort is worth it. Without a trace of light pollution in the sky, the only limit on what you see is caused by your eyesight. Someone with perfect vision and who is dark adapted in a site free from light pollution can see more than 5,000 stars, and the Milky Way may even cast a shadow on the ground!

In Chapter 17, I look at some of the world's darkest sites and what you can expect to see there.

The Bortle Scale of sky brightness

One way of estimating the darkness of your night sky is to use the *Bortle Scale,* a nine-point scale of sky quality running from 1 (excellent dark-sky sites) to 9 (the brightest inner-city skies). Table 2-1 lists what each rating means.

Table 2-1	The Bortle Scale
Bortle Rating	*What It Means*
1	Excellent dark-sky site
2	Typical truly dark site
3	Rural sky
4	Rural–suburban transition
5	Suburban sky
6	Bright suburban sky
7	Suburban–urban transition or full Moon
8	City sky
9	Inner-city sky

How good is your observing site?

Maybe you have two different stargazing sites in mind and want to figure out which one will work best. Or perhaps you want to keep a record of how your sky quality changes over time. You can, of course, buy a light meter and take measurements, but a much simpler way of calculating sky quality is to count stars.

The more stars you can see, the darker your sky is, but counting all the stars visible in your sky – even if you're observing from a bright inner city – can be a daunting task. Here's how to make the process simpler:

1. Find the constellation of Orion.

You'll have to wait until Orion is above the horizon to find it. The months of January through March are best because Orion is visible in the early evening, but you should be able to find Orion by staying up late during October through December, too. Orion has lots of stars in it, but the basic shape is made up of four stars in a rectangle, with three fainter stars in a diagonal line in the centre of this rectangle (see Figure 2-1).

Figure 2-1: The best time to observe Orion is January through March, although you'll be able to see it from October through December.

2. After you find Orion, wait for your eyes to adapt to the dark and then count how many stars you can see inside the main rectangle.

This number gives you an indication of how good your sky is. The more stars you can see, the better!

Table 2-2 gives you the approximate number of stars you can expect to count in different stargazing sites.

Table 2-2	Number of Stars in Different Sites
Bortle Rating	*Approximate Number of Stars*
1	Far too many to count!
2	Too many to count!
3	More than 30 stars
4	Around 30 stars
5	Around 20 stars
6	Around 12 stars
7	Around 6 stars
8	Around 3 stars (the belt stars)
9	One or 2 stars

Knowing When to Head Out

The time when you go out stargazing is mainly dictated by what you want to see. If you really want to find a specific constellation, you usually have to wait until the right season. If you want to find a certain planet, then you'll need to know when that object is above the horizon; you may have to get up before sunrise to see it or wait months until it's easily visible again.

Plan your stargazing month by month. Sit down at the start of each month and work out which days, weather permitting, will be the best ones to go out and find the things you want to see.

Star counts

You can participate in many citizen science projects that measure the relative darkness of stargazing sites. These projects include Globe at Night, the Great World Wide Star Count, and in the UK, the Campaign to Protect Rural England's star count week. By participating in one of these star counts, you'll be providing valuable data to allow astronomers to monitor the spread of light pollution and thereby present a stronger case when looking to persuade local and national agencies that the night sky is part of the natural environment and equally worth preserving as the rivers, forests and mountains.

Figuring out what you want to see

Plenty of great resources allow stargazers to work out when to head outside based on what's up in the sky that night. A simple *planisphere* tells you what constellations are up throughout the year, and the good ones tell you where and when the planets are visible. Planispheres are circular star maps, with two pieces. The lower piece has all the stars that are visible from a specific viewing location all year round. The upper piece has a small clear window in it, and by rotating this window and setting it to a specific time and date (using marks around the outside of the planisphere), you can find what objects are overhead on any night of the year. (See Chapter 9 for more on using a planisphere.)

If you're determined to hunt down those faint fuzzies, then you need a more detailed star map. You can find good star maps by subscribing to any of the excellent monthly astronomy magazines, which all have detailed maps of what's up when and focus on what objects are best viewed each month.

Meteor-chasers need to be much more specific about when they head outside. Although you can see some meteors on any good, clear night, you really haven't lived until you've witnessed a meteor shower, when instead of one streak of light every hour or so, you may be lucky enough to see *hundreds* of shooting stars every hour. A meteor shower is one of the most awe-inspiring sights in nature. Table 2-3 lists the dates of the prominent meteor showers.

Table 2-3	Meteor Shower Calendar
Shower	*Date*
Quadrantids	Early January
Lyrids	Mid-April
Perseids	Mid-August
Draconids	Early October
Orionids	Late October
Leonids	Mid-November
Geminids	Mid-December

Although the dates in Table 2-3 are generally accurate, you'll need to find out the exact timings of specific meteor showers; there's no point expecting to see anything when you head outside at 9 p.m. if the shower doesn't start properly until 4 a.m.!

Moon or no Moon?

If you want to see the Moon – or more likely, if you don't want to see it! – then you need to plan your stargazing around the Moon's own calendar, its phases. Over the course of 29 days, the Moon grows from a new Moon, where you can't see it at all, to a full Moon, high in the sky at midnight, and back to a new Moon again.

If the Moon is anything bigger than a thin crescent, it can harm your dark adaptation, especially if you're stargazing from a dark site. Most stargazers avoid going out stargazing when a bright Moon is in the sky. You'll have to become familiar with the dates of the Moon phases and the times it rises and sets. This information is sometimes marked in diaries or calendars, but you can find that data and more online or in your monthly astronomy magazine.

After dark

Star light, star bright, first star I see tonight . . . is probably a planet. After the Sun sets, the sky is still really bright, and you won't be able to see anything except the Moon (if it's above the horizon). After a short while, the sky dims, and you'll start to see the brighter objects emerge from the blue. The first ones you'll see will be the brightest ones, which are usually planets, with Venus and Jupiter shining really brightly if they're above the horizon.

As you wait outside longer and the sky fades from bright to dark blue, more stars will appear, until eventually all the sunlight is gone from the sky and you're left with black overhead (or orange if you live in a town or city).

Most urban or suburban stargazers don't have to worry about waiting until the end of astronomical twilight (see Chapter 1) to go out stargazing; the sky glow overhead is the main limiting factor on what they can see. However, if you're out somewhere really dark, then you should wait until the end of astronomical twilight, at which point the sky is completely free of sunlight and you can get an unspoiled view.

Looking Up for the First Time

So, you're dressed warmly, you've wrapped red cellophane around the front of your torch, you've hunted down your local stargazing spot, and you've given your eyes at least ten minutes to adapt to the dark. You're ready for your first look up! (Okay, so you probably sneaked a peek before your eyes were perfectly dark adapted. That's fine.)

The first thing you'll realise is that an awful lot of stuff is up in the night sky. You'll have seen the stars before, and you may have been able to pick out a constellation or two – the Big Dipper or the Southern Cross, maybe Orion the Hunter – but hopefully, with all your preparation, you'll now see much more.

Don't get lost!

The first thing that stargazers do when looking up is to orient themselves so they know which direction is which, and you'll need to do the same. Observers in the northern hemisphere have an easier task than their southern counterparts, as one very famous star sits directly above the North Pole. If you can find that star, you're looking due north. *Polaris,* or more commonly the North Star, is the star you need to find. The North Star is actually not that bright, so it doesn't stand out among the background of other brighter stars. This means you'll need to use a signpost to find it, and in most cases, you'll use the bright pointer stars of the Big Dipper.

The Big Dipper is a bright prominent *asterism,* or pattern of stars, with the familiar shape of a ladle or big scoop (see Figure 2-2). The two stars farthest from the handle of the Big Dipper are known as the *pointer stars;* if you draw a line from them, it'll point to *Polaris.* Stargazers north of 35°N will always be able to find the pointer stars of the Big Dipper. If you're in the northern hemisphere but south of 35°N, then you'll need to use a different method when the Big Dipper isn't visible. The constellation Cassiopeia is made up of five bright stars in a zigzag, or W shape. Imagine the left-hand V of the W is a mouth, with its opening toward *Polaris*. If you draw a line straight out of the mouth, it will point toward the North Star.

As the sky changes over the course of a night, the stars in the Big Dipper will change position and won't always look like they do in Figure 2-2, but you can still use the pointer stars to find the North Star, no matter what position the Big Dipper is in.

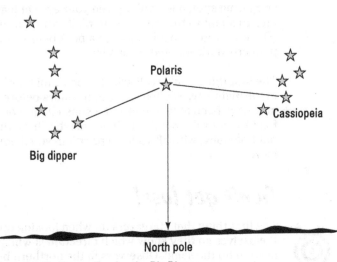

Figure 2-2: Finding north using the Big Dipper.

The southern hemisphere doesn't have a south star, so finding due south takes a bit more practice. The first thing to do is to find your signposts:

1. **Find** the Southern Cross.

   The Southern Cross is a small group of four bright stars in a familiar cross shape, and the southern pointer stars, Alpha and Beta Centauri (see Figure 2-3).

2. **Draw a line along the long axis of the Southern Cross.**

   That line will point towards the South Pole. But to find due south, use the southern pointer stars, too.

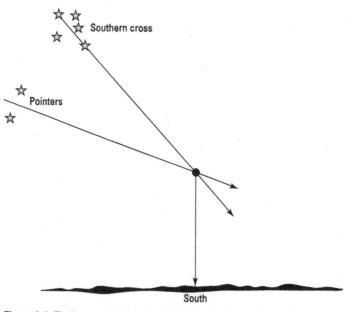

Figure 2-3: Finding south using the Southern Cross.

3. Draw a line that bisects these two bright stars.

Where that line crosses the line drawn from the Southern Cross is where the South Pole star would be. (Remember, a star isn't actually there; you'll just have to pretend!) Stand facing that part of the sky, and you'll be looking south.

Find your signposts

The Big Dipper, Cassiopeia, the Southern Cross and Alpha and Beta Centauri are all very useful pointers, but to make your task easier, you should familiarise yourself with many more celestial signposts.

One of the best signposts in the sky is the constellation of Orion the Hunter. Orion is visible from virtually everywhere on Earth between the months of October and March (although you may have to stay up late to see it if you're observing from October through December).

See Chapter 8 for more information about different signposts in the sky.

Chapter 3

Binocular Astronomy

- -

In This Chapter

▶ Using binoculars for stargazing

▶ Investing in your first pair of stargazing binoculars

▶ Tracking down the great binocular objects in the night sky

- -

*I*f I asked you to name one thing that makes you think of an astronomer, you'd probably say a telescope. And it's true that telescopes can give you stunning views out into deep space and of the planets in the solar system. But telescopes don't do some things very well, and for those, you're better off using binoculars.

In this chapter, I discuss what to look for in binoculars, as well as the many objects you can see with them in the sky.

How Binoculars Work

Most serious amateur stargazers, in addition to a telescope, also have a good pair of binoculars.

Binoculars are:

- ✔ Usually cheaper than telescopes
- ✔ Easier to use
- ✔ Quicker to set up
- ✔ More portable
- ✔ Better for observing certain things
- ✔ Multi-use (you can use them in the daytime, too, for bird-watching, sight-seeing and other things)

Binoculars are simply two small telescopes fixed together, so you can use both eyes to stargaze, which is more natural than the one-eyed squint you have to do when you look through a telescope.

Your binoculars will have two large lenses at the front, which you'll point at the thing you want to see, and two lenses at the back, where you'll put your eyes.

Even though binoculars are easier to use than telescopes, you may initially have trouble tracking down the faint fuzzies that you're looking for. That's because using binoculars at night is a whole different ball game than using them in the daytime.

Lenses and prisms

The large lenses at the front of your pair of binoculars are called the *objective lenses*. The smaller lenses that you look through are called the *eyepiece lenses*. But the objective and eyepiece lenses, which are shown in Figure 3-1, aren't the only bits of glass inside your binoculars.

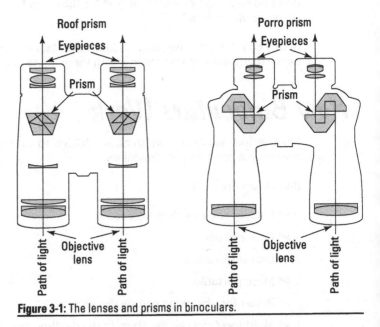

Figure 3-1: The lenses and prisms in binoculars.

Small glass prisms are inside the tubes that make up each side of the binoculars.

The lenses are what focus and magnify your image, while the prisms are there to make sure that the light that comes in through the objective lens comes out of the eyepiece lenses.

Getting focused

If you just pick up a pair of binoculars and look through them, chances are what you're seeing isn't in focus; it'll look blurry and not sharp. Lack of focus is an even bigger problem when you're using binoculars for stargazing. When stars are out of focus, their light gets smeared out into disks which are much harder to see than the sharp bright dots of the stars. Turning an unfocused pair of binoculars up to the night sky may make you think that no stars are there at all.

Get your binoculars focused before you try to find anything in the sky. Once your binoculars are focused, the stars you see should look incredibly sharp.

You can focus most binoculars in the same way. Just follow this simple step-by-step guide, shown in Figure 3-2:

1. **Adjust the separation between the two eyepiece lenses so that they're the same spacing as the distance between your pupils, called the interpupillary distance.**

    You can usually adjust the separation by holding both halves of the binoculars and physically twisting them apart or together.

2. **Point your binoculars at something bright but distant.**

    A planet or a bright star like *Sirius* or *Canopus* (see Chapters 11 and 12) works well.

3. **Have a quick look-through.**

    You never know; they may already be in focus!

4. **Close the eye that looks through the eyepiece with an independent focus.**

    You may find it easier to put the eyepiece lens-cap back on so that you don't have to squint.

5. **Adjust the focus until your target looks sharp for your left eye.**

   Use the focus ring normally found at the top of the binoculars on the *bridge* between both halves.

6. **Close your left eye and open your right eye (or swap the eyepiece cap onto the other side) and adjust the focus for your right eye using the eyepiece focus, called the dioptre corrector.**

   The dioptre corrector is for people who have two eyes of slightly different strengths.

You're now good to go, but remember that everyone's eyes are different. If someone else uses your binoculars, you may have to repeat the focusing process to get them back in focus for your eyes.

Some binoculars have different focus adjustments. Some may have only two eyepiece focus rings, and some may not have any focus adjustments whatsoever. These latter models of binoculars are designed for people with good eyesight, but don't allow you to correct for less-than-perfect eyesight. In general, you'll want to focus your binoculars one way or another.

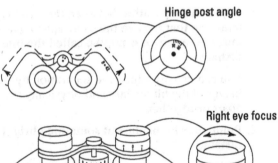

Hinge post angle

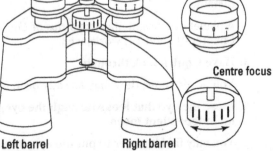

Right eye focus

Centre focus

Left barrel   Right barrel

Figure 3-2: Follow these simple steps to adjust your focus.

If you wear spectacles, you can decide whether it's easier to leave them on or take them off when you're using binoculars. I usually take mine off, because I find this more comfortable. Your binoculars will work fine if you leave your glasses on, and you won't have to worry about losing your specs in the dark.

Figuring Out Which Binoculars to Buy

Because such a huge range of binoculars are available, picking the right pair may be a complicated process. Luckily, you can look for a few simple things that will help you narrow the field and find the best binoculars for you.

 Binoculars have a simple rating system that allows you to work out exactly how clear, bright and large an image you can expect to see through them.

Power matters

Binoculars' vital statistics are quoted using two numbers separated by a multiplication sign – for example, 10×50, pronounced '10 by 50'. These numbers tell you the magnifying power and the light-gathering power of the binoculars, important information for you as a stargazer.

The first number listed (the '$10 \times$' bit) refers to the *magnifying power* of the binoculars, so a pair of 10×50 binoculars magnify things ten times. The greater your magnifying power – that is, the higher the first number – the bigger everything will appear when you look through the eyepieces.

 Despite what you may think, having a high magnifying power isn't the most important thing. In fact, the higher the magnifying power, the dimmer the object will appear. Although you're still collecting only the same amount of light, it's now spread out over a larger image, which is why lower-power binoculars in general give brighter (albeit smaller) images.

The second number listed (the '50' bit) tells you the diameter, in millimetres, of the front objective lenses. This number is referred to as the *light-gathering power*. The larger this number, the brighter things will look, because you're physically gathering more light to pack into the small image.

It's not the diameter so much as the area of the objective lenses that's important, and the area is proportional to the diameter *squared.* If you double the objective lens diameter, you'll increase the light-gathering power by a factor of four.

However, binoculars with large objective lenses and high light-gathering power have two main drawbacks:

✔ They are more expensive than binoculars with smaller objective lenses.

✔ They are much heavier and therefore more difficult to hold up for long periods of time stargazing.

As with most things, you'll need to compromise when buying a pair of binoculars. Yes, you'd like huge objective lenses combined with a high magnifying power, but such binoculars are big and heavy (and expensive).

For example, a giant pair of 25×100 stargazing binoculars will give you images two-and-a-half times bigger than a pair of 10×50s and will gather four times as much light. But they'll be much more expensive and significantly heavier. A normal pair of 10×50s will weight around 1kg, whereas a pair of 25×100s will weight upwards of 3kg, and you won't have to be holding those aloft for very long before your arms start to ache.

Field of view

The *field of view* (FOV) of your binoculars is the area of sky that you can see using them. The more powerful your magnifying power is, the smaller the patch of sky you'll see. And, in some cases, you'll want a large FOV – and therefore a smaller magnifying power – to observe some of the largest faint fuzzies. It's worth checking what your binoculars' FOV will let you see before you buy them.

Astronomers measure distances in the sky in terms of degrees – the angular distance between two objects. As a general rule, your clenched fist held at arm's length is around ten degrees tall. Your binocular FOV will be stated in degrees.

The angular distance between the two brightest stars in Orion –
Betelgeuse and *Rigel* – is 18°, so you can work out from that
how much of Orion your binoculars might see. Figure 3-3
shows you the possible FOVs of a variety of different-powered
binoculars.

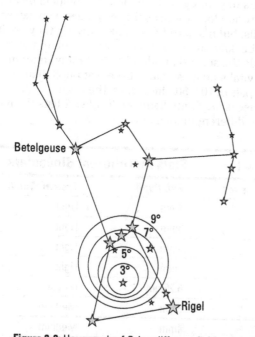

Figure 3-3: How much of Orion different fields of view allow you to see.

Exit pupils

There's another important factor for you to consider when
buying stargazing binoculars, and it's related to your age and
consequently how youthful your eyes are!

The *pupil* – the black bit at the centre of your iris – is where
light gets into your eyes. In bright sunlight your pupils shrink
down to a diameter of around 1.5mm, but in dim light they
grow to up to 8mm diameter. However, your pupils' ability to
expand in dim light deteriorates as you age. Young stargazers'
pupils can open up to the full 8mm, while the pupils of those
in their 70s may only open to 3mm! How much your pupil
expands affects how much light goes into your eye, and so

older stargazers will see things looking much less bright and clear than young stargazers do.

The beams of light that come out of the binoculars' eyepiece lenses are called the *exit pupils,* and you'll want to match the binoculars to your eyes. If you hold the binoculars away from your body and look towards the eyepieces, you can see the exit pupils, but it's hard to estimate how big they are just by looking. Luckily, there's a handy rule for you to remember. If you divide the second number by the first number in your binoculars' vital statistics, you get the exit pupil diameter in mm. So, for a pair of 10×50 binoculars, the beams of light leaving the eyepieces are 5mm diameter. Table 3-1 lists the measurements for different binoculars.

Table 3-1	Stats for Different Binoculars	
Vital Statistics	*Exit Pupil*	*Typical Weight*
8×32	4mm	Light
8×40	5mm	Light
10×30	3mm	Light
7×50	7mm	Light
8×50	6.25mm	Medium
10×50	5mm	Medium
12×50	4mm	Medium
16×50	3mm	Heavy
20×60	3mm	Heavy
15×70	4.67mm	Heavy
25×100	4mm	Very heavy

Eye relief

Another important stat to bear in mind when you're looking to buy a pair of binoculars is the *eye relief.* Eye relief is the distance at which you have to hold your eyes from the eyepiece lenses to see an image. If a pair of binoculars has a very small eye relief, you may find yourself having to almost touch the eyepiece lenses with your eyeball – never a good thing!

Look for binoculars with a decent eye relief, especially if you wear glasses and don't want to have to take them off every time you want to look through your binoculars.

Lens coatings

Some more expensive binoculars have coatings on the lenses that allow more light to get through and therefore give you a brighter image. Whenever light passes through glass, which happens in the multiple lenses and prisms in your binoculars, some of that light gets reflected back the way it came. In some very cheap binoculars, as much as 50 per cent or more of the light can reflect back out, so you'll see a much dimmer image than you may expect.

Lens coatings can greatly improve the amount of light that gets through to your eyes, but they'll also add a hefty sum to the price tag. Paying extra, though, may be worth it, because the images will look much brighter and have better contrast than anything you'll see through similarly sized uncoated binoculars.

Using a Steady Hand or a Tripod

The first time you take your binoculars out stargazing, you may be disappointed to find that all the stars, planets and faint fuzzies that you'd planned on observing are dancing all over the place when you look through the eyepieces. Don't worry, the universe isn't moving; it's just you.

No matter how still you try to hold your binoculars, small wobbles in your arm, hands or head, and even your breathing, will cause the stars to move to such an extent that you may struggle to see much. So you need to find a way of steadying your binoculars to reduce the wobble.

You can try balancing your binoculars on a nearby wall or perhaps your car roof or some other solid surface, but even these steady surfaces won't cure the view of all wobbles. You'll still have to hold your binoculars with your hands, and that means you're the limiting factor.

Some high-tech binoculars have image-stabilising hardware that reduces the wobble, giving you a stiller image. Image-stabilised (IS) binoculars aren't cheap. However, you may be able to find a small pair of IS binoculars that give you a better view than a non-IS pair. If you can afford a lower-power set, which saves a bit of money over a high-power set, that's the way to go.

Tripods: Three legs to stand on

Tripods are great additions to your stargazing kit, especially if you're using binoculars. By attaching your binoculars to a tripod, you instantly steady the image, because you don't have to touch the binoculars at all. You can't attach all binoculars to a tripod, though, so you need to find out whether you'll be able to fix yours on or not.

Some binoculars have a screw thread with which you can use an adaptor to attach them to a tripod. To find this screw thread, you may need to remove a small plastic covering cap, which will be hiding it from view. Most often, you can find the screw thread in the central bar of the telescope at the front, between the two objective lenses.

If your binoculars don't have a screw thread to attach them to a tripod, you may be able to use an adaptor that clamps onto the pivot bar between both barrels of the binoculars.

If the surface your tripod is standing on isn't solid, or if it's windy, you still may have a bit of a wobble. The best place for binoculars on tripods is somewhere sheltered that's firm underfoot.

Monopods: One leg to stand on

In some cases, a tripod can be a bit awkward to manoeuvre around, especially if you're planning on observing lots of different targets in the sky. The next best thing to a tripod is a *monopod,* which is really just a one-legged tripod.

'But won't it fall over?' I hear you ask. No, not if you hold onto it! But if you're holding the monopod, you may wobble and cause the image to jump about, and keeping the monopod steady is far more difficult that keeping a tripod steady.

Having said that, monopods do have their uses. They:

- Are lighter and more portable than tripods
- Allow you to take the weight of your binoculars off your arms
- Steady the image more than you can just by holding the binoculars
- Make it more easy to pivot the binoculars around to look at many different objects

Getting comfy using a tripod or monopod

After you've mounted your binoculars on a tripod or monopod, you're almost ready to start some serious stargazing! One final thing you'll have to consider is how to get comfortable looking through the eyepieces.

Most tripods won't extend high enough to raise the eyepieces to a comfortable height, so you'll have to stoop. And if the thing you're looking at is high overhead, you may have to simultaneously stoop and crane your neck, which isn't exactly a comfortable position to be in.

One of the best ways to overcome this awkwardness is to stargaze while sitting or lying on a reclining outdoor chair. Now that's stargazing in comfort! This position is where a monopod comes into its own, as it's much easier to move around than a tripod.

If you're stargazing from a chair, make sure that you dress warmly and put a blanket between yourself and the chair for insulation. If you're not moving about much, you can cool down pretty quickly.

A Binocular Bonanza

A wealth of astronomical objects look better through binoculars than through a telescope. Although telescopes usually have higher magnifications and light-gathering powers compared with binoculars, their field of view is much smaller, so you'll be looking at a small patch of sky, and maybe only a tiny corner of a faint fuzzy.

You'll need to get away from light-polluted skies to get the best views of the night sky, and faint fuzzies will stand out much more clearly against a black sky than against a lit-up orange one. However, if you're stargazing with binoculars in a town or a city, you'll still see more than your eye alone will see, but it may take you a little longer to find things.

The following sections offer a list of some of the best binocular objects in the night sky. For lots more detail on these objects, see Chapters 6 and 7.

The solar system up close

The solar system has a lot to show the binocular stargazer, from craters on the Moon to the rings of Saturn, from Jupiter's largest moons to spectacular comets. Unlike a telescope, your binoculars will magnify things only slightly, so most of the planets don't look like disks; they'll just look like very bright and steady stars. The closest planet to Earth (Venus) and the biggest planet in the solar system (Jupiter) can show as a disk, and you'll see the rings of Saturn too!

The Moon

The Moon looks great through a pair of binoculars, filling the field of view and showing off lots of craters and dark, smooth seas or *mare* (pronounced mah-ray). If you want to explore in detail, see Chapter 7 for a Moon map.

Mercury

Mercury is very hard to spot because it's never very far from the Sun in the sky. Mercury either sets just after the Sun does or rises just before it. Through binoculars, you won't see anything other than a bright-ish star glowing in the twilight, but at least you'll have seen it, which is more than most people can say.

It's very dangerous to look at the Sun through binoculars – it can blind you – so make sure you're not scanning the sky for Mercury when the Sun is up.

Venus

Venus is the closest planet to Earth and so looks biggest in the sky. Still, Venus is very small, and you may struggle to see anything other than a very bright steady star when observing Venus, unless you use medium or bigger binoculars.

What Venus will show you is phases like the Moon – sometimes a half Venus and, best of all, a large, thin crescent Venus. The thin crescent Venus happens when Venus is at its closest to Earth, lying almost between the Earth and Sun, and so it will look larger than it does in the other phases; this is, therefore, probably the best time to try and see Venus through binoculars. Even so, you'll still need a magnifying power of at least 10x, a steady tripod, and very good eyesight to see Venus as anything other than a bright dot.

Mars

Mars is a small planet, and you won't see anything other than a bright dot through binoculars. However, Mars is still worth a look, because you'll see its red colour much more easily through binoculars than with your naked eyes.

Jupiter

Jupiter is the largest of all the planets, which means that even though it's a long way away, you may still be able to make out a disk with a good pair of steady binoculars.

It's not the planet Jupiter itself that's the interesting thing in your binoculars' field of view. Instead, look for up to four tiny little dots lined up with Jupiter; these dots are Jupiter's four largest moons: Io, Europa, Ganymede and Callisto. Sometimes you'll see two on one side of Jupiter and one on the other; sometimes three and one; sometimes four and none; and sometimes you won't see all four, because one or more may be hidden behind or in front of Jupiter.

If you watch Jupiter's moons through binoculars from night to night and draw their relative positions, you'll quickly notice that they don't stay in the same place; you're watching them orbit Jupiter! As long as your binoculars are mounted on a tripod, you should be able to see Jupiter's largest moons with even a pair of 7×50 binoculars.

Saturn

Saturn's rings are something that any budding stargazer wants to see, and with binoculars, you can just about start to make them out.

Looking through a pair of 7 × 50 binoculars, you may notice that Saturn doesn't look quite right; it's slightly elongated. You may even make out small bulges on either side of the disk; they're the famous rings! The larger the binoculars you use, the more distinct these bulges become, but you'll still struggle to see the rings distinctly unless you use a telescope.

Comets

Bright comets are objects that can look far better through binoculars than through a telescope, because they cover a large area of the sky and telescopes would zoom in too much on just one small part of the comet. Some comets are predictable – for example, Halley's Comet, which is due to return to the skies in 2061 – but most bright comets appear out of the blue.

If you ever hear of a comet brightening the sky, make sure you have a look through your binoculars; you won't be disappointed.

The faint fuzzies up close

Some of the best sights you'll see through your binoculars will be the faint fuzzy nebulae that pepper the night sky. You may be able to pick a few of them out with your eyes, but binoculars will show them looking much bigger and more defined.

Here's a list of good binocular targets, but you'll find lots more to look out for in the constellation guides in Part III.

Gas clouds

Gas clouds are, just as the name suggests, large clouds of gas in space – either clouds where stars are being born, or the remnants of a star once it's exploded in its death throes (see Chapter 6).

The **Orion Nebula, M42** (see Chapter 11), is without a doubt one of the best binocular faint fuzzies in the night sky. You can easily see it with the naked eye from a dark site. If you look at it through binoculars, though, you'll be able to see its structure: a grey, nebulous cloud of gas in space. From light-polluted skies you may just be able to make out this fuzziness, but in dark skies with large astronomical binoculars you may be able to make out some gaseous 'arms' extending out from the main bulk of M42.

The **Eta Carinae Nebula, NGC 3372** (see Chapter 12), is bigger and brighter than the Orion Nebula, but because it's visible only in the southern hemisphere, it's not nearly as well known. If you manage to track down this gas cloud with binoculars, you'll see a large fuzzy cloud of gas in the sky. If you're observing under dark skies, you may make out some structure within the fuzziness – in particular, a distinct L-shaped dark band running through it.

Galaxies

Galaxies are huge collections of stars. All the stars you see in the night sky are in our galaxy, the Milky Way, but there are many other galaxies beyond ours, and some of the bigger and closer ones are visible through binoculars.

The **Andromeda Galaxy, M31** (see Chapter 14), is the closest big galaxy to ours, and contains around a trillion stars. Even still, it's very far away, and so only appears as a faint smudge in the sky. Through 10 × 50 binoculars you might be able to enlarge this smudge and see the bright central core of this galaxy, but with larger binoculars, you can begin to see its fainter outer arms.

The **Large and Small Magellanic Clouds** (see Chapter 15) are nearby dwarf galaxies. They're physically much smaller than the Andromeda Galaxy, but are significantly closer and so appear larger in our skies. They're only visible from quite high southern latitudes though. Binoculars won't show the individual stars in these dwarf galaxies, but will show subtle variations in brightness.

Open clusters

Open clusters are groups of stars within our galaxy that all formed out of the same gas cloud.

The **Pleiades, M45,** or the **Seven Sisters** (see Chapter 11), is another target that looks better through binoculars than through telescopes, which usually zoom in too much. Its name suggests there are only seven stars are in this cluster, but through binoculars, you'll see dozens of sparkling stars.

The **Southern Pleiades, IC 2602** (see Chapter 12), is an open cluster, just like its bigger and brighter northern namesake. It won't look quite as bright as M45, but through a pair of 10 × 50 binoculars you'll see plenty of stars. It's not visible to most northern hemisphere stargazers though.

The **Double Cluster in Perseus, NGC 869 & 884** (see Chapter 14), looks beautiful through even a small pair of binoculars. The two side-by-side star clusters make it unmistakeable, and you can easily find it with the naked eye, lying half way between the bright constellations Cassiopeia and Perseus.

Globular clusters

Globular clusters are spherical collections of stars that lie in a halo around our galaxy. Some of the most noteworthy are:

- ✓ **Omega Centauri, NGC 5139** (see Chapter 12): The best globular cluster in the sky. Through a pair of 10 × 50 binoculars, you'll be able to tell that it's not a sharp star like the others around it; it's got a distinct globular faint fuzziness, hence the name *globular cluster*. It's not visible to most northern hemisphere stargazers though.

- ✓ **47 Tucanae, NGC 104** (see Chapter 15): Another good globular cluster, not quite as good at Omega Centauri, but better than the Great Hercules Cluster. Like Omega Centauri, though, it's a southern hemisphere object.

- ✓ **The Great Hercules Cluster, M13** (see Chapter 13): One of the best globular clusters in the northern skies, but it doesn't look quite as good as the best of all globular clusters, Omega Centauri. If you have good dark skies and a pair of 10 × 50 binoculars, you should be able to tell that it's not star-like; it's a tiny elongated blob. Bigger binoculars will let you see the fuzziness more clearly, but even still, it will remain pretty small in your field of view.

Double stars

Although stars in the sky look like single dots to our eyes, many of them are in fact double stars – that is, two stars side by side in space, or maybe just along the same line of sight. Our eyes aren't good enough to make out the two individual stars, but binoculars can help.

Mizar and **Alcor**, in the constellation of Ursa Major (see Chapter 10), are among the best double stars in the sky. With just your eye you may be able to tell that more than just one star is here, but using binoculars you'll see them really clearly. If you're using a pair of large astronomical binoculars with a magnification of 10x or more mounted on a tripod, you may also notice that the brighter of this pair, *Mizar*, is itself a double star.

Albireo is a double star in the constellation of Cygnus the Swan (see Chapter 13). You can easily separate *Albireo* into its two components. Through high-power binoculars (20x or more), you'll notice that these two stars are different colours. The fainter of the pair is blue, and the brighter is orange.

Chapter 4

Your First Telescope

- -

In This Chapter

▶ Choosing your first telescope

▶ Setting up your telescope

▶ Tracking down great telescope objects in the sky

- -

*F*or most people, stargazing means using a telescope,
but as you can see in Chapters 2 and 3, you can view a
lot with just your naked eyes or binoculars. However, some
things in the night sky definitely require a telescope to see,
and most of your unaided-eye and binocular targets will look
much better through a telescope, too.

In this chapter, I look at the different types of telescope available and help you work out which is the best telescope for
you. After you buy your first telescope, you need to practise
setting it up until it's second nature. Nothing will put you off
stargazing more than wrestling with your unwieldy telescope.
After you have the hang of how to use your telescope, the
fun begins and you'll want to set your telescope up at every
opportunity to chase down those great stargazing sights.

Deciding on a Telescope

Telescopes come in a baffling array of shapes and sizes, which
makes finding the right one for you a little tricky. Luckily,
you can follow a few easy rules to make sure you choose the
right one.

If you're new to stargazing, you may want to consider starting
with a good pair of binoculars (see Chapter 3). Binoculars are
cheaper, easier to use and more portable than a telescope,
making them perfect for beginners. After you've exhausted

your list of binocular targets, then you can upgrade to a
telescope.

Reflectors versus refractors

Start your decision by looking at the different models of telescope tube – the bits with the optics in. You can find quite
a few different designs. *Reflecting telescopes* use mirrors to
gather the light. *Refracting telescopes* use lenses. There are
different kinds of reflectors, but in general the refractors all
follow the same basic design.

Refractors

Refracting telescopes use lenses to collect and focus the light,
just like binoculars do. In fact, you can think of a refracting
telescope as one half of a giant pair of binoculars.

The light enters a refracting telescope through the front lens,
called the *objective lens* (see Figure 4-1). It then travels down
the length of the telescope to the eyepiece lenses, which is
where the magnifying happens.

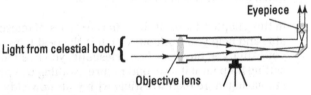

Figure 4-1: A refracting telescope.

This setup can make observing through a refracting telescope
quite uncomfortable, especially if you're pointing your telescope high in the sky, because you may have to stoop and
crane your neck to see through the eyepiece. Some telescopes
come with a prism adaptor that bends the light through
90 degrees, so looking through the eyepiece is much more
comfortable.

Refracting telescopes have several benefits over other
telescopes:

  ✔ Because the inside of the tube is sealed at both ends
    (with lenses), refracting telescopes don't suffer from
    dirt inside.

✔ Because the tube is sealed at both ends, refracting telescopes don't have the problem of air moving about inside the tube, and so have sharper, steadier images.

However, refracting telescopes are longer and more unwieldy than reflecting telescopes, and can suffer from something called *chromatic aberration,* where a rainbow of colours appears around the image. Lens coatings and a longer telescope tube (which increases something called the focal length) can help reduce this problem.

Reflecting telescopes

One of the most common types of reflecting telescope is called a *Newtonian reflector,* named after the man who invented it, Isaac Newton (of falling apple fame).

In Newtonian telescopes (see Figure 4-2), the tube is open at one end. The light enters the tube and reflects off a curved mirror (called the *primary mirror*) at the other end, before bouncing back up the tube to near the top where it reflects off a smaller mirror (called the *secondary mirror*), which bounces the light out of a hole in the side of the telescope where you attach your eyepiece.

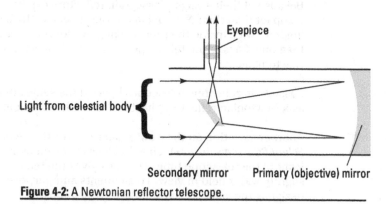

Figure 4-2: A Newtonian reflector telescope.

This setup can make looking through the eyepiece a little tricky, especially in the bigger Newtonian telescopes, where you may need a step or small ladder to get high up enough to see through the eyepiece.

The benefits of Newtonian reflecting telescopes over other telescopes include the following:

✔ Because the mirror can be fixed onto a metal plate, reflecting telescopes can be much bigger than refractors.

✔ Reflecting telescopes are cheaper to make.

✔ Reflecting telescopes don't suffer from *chromatic aberration.*

A variation on the standard Newtonian telescope is the *Cassegrain reflector,* shown in Figure 4-3, which uses a curved secondary mirror to bounce the light back down the length of the tube to an eyepiece at the bottom end.

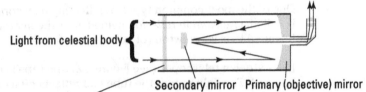

Light from celestial body

Secondary mirror   Primary (objective) mirror

Correcting lens
(in Schmidt-Cassegrain telescope)

Figure 4-3: A Cassegrain reflector telescope.

Because of their design, Cassegrain reflectors can be more compact than their Newtonian cousins, because the light is 'folded' twice inside the tube, making two journeys rather than one. Cassegrain telescopes can be half the length of Newtonians.

Cassegrain reflecting telescopes have all the same advantages as a Newtonian reflector, plus they're much more compact.

Another variation on the Newtonian reflector theme is the *SchmidtCassegrain,* in which a thin lens is placed over the front of the telescope tube. This lens gives the telescope a slightly wider field of view. To all intents and purposes, the design is the same as the standard Cassegrain.

Mounts

As if it isn't bad enough that you have so many pros and cons to consider when buying a telescope tube, you also have a wide range of telescope mounts to choose from, too.

Focal length and focal ratio

You'll often see the term 'focal length' relating to telescopes. The focal length of a telescope is the distance (usually quoted in millimetres) between the primary lens (or mirror, in the case of a reflecting telescope) and the point where the light comes to focus. The longer the focal length, the longer the telescope tube, but in the case of a Cassegrain reflector telescope, the light is 'folded' inside the tube (it's bounced back and forth), so the length of the telescope tube is far less than the focal length of the telescope.

Another important factor in telescopes is the focal ratio, which is calculated by dividing the focal length by the aperture diameter. For example, a telescope with an aperture of 200mm and a focal length of 2000m would have a focal ratio of 10, written f/10. The lower the focal ratio, the wider the telescope's field of view.

Alt-az

The most common kind of mount is the alt-az mount, which is short for altitude-azimuth. These mounts don't look very complicated; they simply allow you to tilt the telescope up and down (changing the altitude) or from side to side (changing the azimuth), in much the same way that a camera tripod works. This arrangement makes using alt-az mounts relatively straightforward; just point the telescope at what you want to look at and away you go.

Alt-az mounts are easy to set up and use.

Equatorial

If you're going to be observing for hours at a time, maybe hoping to track the same object as it moves across the sky, then an equatorial mount may be for you. These mounts allow you to move the telescope around two axes, and just like when using the alt-az mount (see preceding section), you can move the telescope up and down or from side to side. However, on an equatorial mount, you're moving the telescope through two new coordinates: *right ascension* (RA, represented by the Greek letter alpha, α) and *declination* (dec, represented by the Greek letter delta, δ). See Chapter 9 for more on RA and dec.

On an equatorial mount, your whole setup is tilted backwards. The tilt of the mount is equal to your latitude in degrees. As a result, as you track an object across the sky, you have to move only one axis of the mount, the RA axis. Many equatorial mounts are motor driven, meaning they'll track the object for you, freeing you to simply stargaze.

The benefits of the equatorial mount include the following:

- An equatorial mount can track the stars by moving only one axis.
- An equatorial mount is easily motorised to allow automatic tracking.

Dobsonian

Perhaps one of the best mounts for new stargazers is the *Dobsonian mount*. John Dobson designed this cheap and simple mount in the 1950s, enabling stargazers to save money on the mount and afford a bigger telescope!

Dobsonian mounts usually house larger telescopes. Rather than using a sophisticated tracking system like the equatorial mounts do, the Dobsonian is essentially just a spinning disk onto which the telescope is fixed in a cradle. To move the telescope around, you only have to spin the disk and move the telescope up and down in the cradle. In this respect, Dobsonian mounts are essentially alt-az mounts.

The benefits of the Dobsonian mount are that:

- They are the easiest mounts to use.
- They are very robust.
- They are very cheap to make.
- They can support a heavy telescope.

One major disadvantage, though, is that Dobsonian telescopes tend not to have tracking mechanisms, so you constantly have to nudge them to follow your target across the sky.

(Aperture) size matters

When stargazing, size matters! The bigger your telescope, the more detail you'll see, the brighter objects will appear, and the clearer they will be. The size you're interested in is the diameter of the barrel of the telescope, called the *aperture.*

Apertures are usually quoted in inches or centimetres and give you a very ready way to work out how good the telescope is.

The aperture isn't the only thing that's important, but it is one of the most important factors. If you can stretch your budget to a bigger telescope, you'll see so much more.

As the aperture increases, your telescope's *light-gathering* power also increases. The light-gathering power is proportional to the area of the barrel that you point up to the sky. If you have two telescopes, one with twice the aperture of the other, your larger telescope will gather four times more light, and so images will be four times brighter and clearer.

In the rest of this book, I refer to small, medium and large telescopes. In general, small telescopes have an aperture of 10cm (4 inches) or less; medium telescopes are between 10 and 20cm (4 and 8 inches); and larger telescopes have apertures wider than 20cm (8 inches).

Light-gathering power

The light-gathering power of a telescope increases with the area of the mirror or lens that's gathering the light. Here's a list of telescope apertures along with their comparative light-gathering power:

Aperture	Light-gathering Power
5cm	1
7.5cm	2.25
10cm	4
15cm	9
20cm	16
25cm	25
30cm	36

Portability versus power

The bigger the telescope you get, the more you'll see; things will look bigger and brighter through a larger telescope. But don't rush out and buy the biggest one you can find. You first need to think about how you'll be using your telescope. Will you be carrying it in and out of your house every night? Will you have to pack it into your car and drive somewhere dark to use it?

If you're moving your telescope around much, you may want to consider portability as an important factor. There's no point buying a 30cm (12-inch) Dobsonian reflector if it's too big and heavy for you to move around a lot.

Manual or automatic?

Depending on what you're planning to use the telescope for, you may want to consider getting one with automatic star tracking or even one with an onboard *go-to* computer, which will find everything for you.

These fancy mounts will put the cost of your telescope up, and you'll need to make sure they're charged when you want to use the telescope, or that you have some means of getting power into them. You do this with a rechargeable power pack, which you'll need to remember to charge before you go out stargazing.

In order for these automated systems to work well, you have to spend quite a bit of time setting up the telescope, ensuring that it's correctly aligned and inputting a few bright target stars so that the telescope's go-to system knows where it's pointing. This process takes a bit more time than just placing a manual telescope on the ground, but despite this you'll soon realise the value of an automated system, because it frees you to do what you wanted to do in the first place – stargaze. Some very sophisticated new telescopes have an automatic setup system, removing any hassle!

The shake test

I discuss a variety of mounts and their relative merits earlier in this chapter, but there's no substitute for seeing the telescope and mount that you plan to buy set up, allowing you to perform one of the most important tests of your new telescope: the shake test.

In general, you want your telescope and mount to be as solid and robust as possible. If your telescope shakes and wobbles when you look through it, you'll never see a very good image and may quickly get frustrated and give up.

 When you're perusing telescopes to decide which to buy, make sure that you see each one set up on the mount, place your hand gently on top and give it a shake. If the telescope moves much, if it feels a bit too loose or if the mount is too light and easily moved, then avoid buying it. You should look for a telescope and mount that barely moves when you touch it, which will give you hours of satisfying stargazing compared with hours of frustration when using a shaky setup.

Eyepieces

Any good new telescope will come with a selection of eyepieces that you can attach to the telescope's eyepiece holder. These little lenses provide the magnification. By swapping different-powered lenses in and out, you can increase or decrease the magnifying power of your telescope.

Eyepieces always have a number written on them, which is the *focal length* in millimetres. The smaller the number, the more the eyepiece will magnify an image.

 You'll usually want to start your evening's stargazing using a wide-field, low magnification eyepiece to find the objects you want to observe. After you find them, you can switch to higher magnification (lower focal length) eyepieces to zoom in.

Eyepieces are a really important part of your telescope kit, so keep each one safe when you're not using it, in its protective box or plastic sleeve. A scratched and scuffed eyepiece doesn't show nearly as much as a clear new one.

If you're looking to upgrade your telescope to a new one, consider investing first in a few new eyepieces. You may be surprised at how much better they make your old telescope.

Having a range of eyepieces – from high-power 5mm lenses down to low-power 25mm lenses – will help you explore the sky at a variety of magnifications and see so much more than you would using just one or two eyepieces.

Magnifying power

To work out the magnifying power of your telescope, you need to divide the focal length of your telescope (usually written on it somewhere) by the focal length of the eyepiece you're using.

For example, a telescope with a 1000mm focal length and a 20mm focal length eyepiece attached will magnify things 50× (1000 divided by 20). If you then replace the 20mm eyepiece with a 5mm one, your magnification jumps up to 200× (1000 divided by 5).

But magnifying power isn't everything. No matter how much you magnify an object, the amount of light you receive from that object stays the same. As you step up in magnification, you lose brightness because you're spreading the same amount of light over a larger area. Don't go straight for the high-power eyepieces; you may get a better view using one with a lower power. Far more important is the light-gathering power of the telescope (the diameter of the aperture).

Field of view

As you change eyepieces in and out to get different magnifications, you'll also get different *fields of view*. The field of view of a telescope is the area of sky you can see when looking through the eyepiece. The more powerful magnifications you use, the more the telescope will zoom in on a part of the sky, and so the smaller your field of view will be. In some cases, you want a wide field of view; if you zoom in too much, you may miss part of the faint fuzzy you were looking for.

Locating anything using a low-power eyepiece with a large field of view is much easier than with a high-power eyepiece with a small field of view. After you find what you were looking for, you can centre it in the middle of the field of view and use more powerful eyepieces to zoom in.

Barlow lenses

Some telescopes come with an extra lens, called the *Barlow lens,* that doesn't quite look like the others. In fact, you can fix your other eyepieces in one end of the Barlow lens and then slot the whole thing into the eyepiece tube of your telescope. Barlow lenses increase the magnification even more (usually doubling it), so with two eyepieces and a Barlow lens, you effectively have four different magnifications.

Features to avoid

So long as you know what you're buying, there's no such thing as a bad telescope. If you're on a limited budget, then a cheap, light telescope may be just the thing for you, allowing you to explore the Moon's surface and a few other bright objects.

If you're planning to spend a bit more on a telescope, then you can start to overcome some of the problems associated with cheaper models. Even on a budget, you'll want to avoid telescopes that have:

- ✔ **Plastic lenses:** Plastic lenses are cheap, and so you sometimes find them in very small, cheap telescopes. Plastic lenses don't give nearly as clear an image as glass lenses do.

- ✔ **Too-high magnifying powers:** You may also find small, cheap telescopes advertising themselves by their magnifying power (for example, 100× magnification!), but that number alone isn't important. You want a good balance between light-gathering power and magnification, so a larger telescope with lower magnification is usually far more useful than a small telescope with high magnification. (For more on this topic, see the earlier section 'Magnifying power'.)

- ✔ **Shaky mounts:** In short, avoid them. (See the earlier section 'The shake test'.)

- ✔ **Poor build quality:** You'll also want to check the whole telescope for build quality. If any part of your telescope is very low quality, then you'll end up getting frustrated with it when it doesn't perform as you'd expect it to.

Storing Your Telescope

Telescopes are fairly simple devices – really just long tubes with either lenses or mirrors inside – but you'll still need to care for them properly when they're not in use. Having somewhere safe to store them is important, because any knock, bump or dent may damage the glass or knock the mirrors or lenses out of alignment, making the telescope next to useless.

Keeping your telescope dry and away from excessive condensation and moisture is important, too. Any buildup of water on the lenses or mirrors may mark their glass, making them far less effective. Many keen astronomers, after they've invested in a large telescope, start thinking about building or buying an observatory – a building with a rotating or removable roof where you can store your telescope safely in the daytime and use it at night without worrying about carrying it outside and setting it up.

You can buy off-the-shelf observatory domes in a variety of sizes, as well as flat roll-off-roof observatories. Before you buy an observatory or start building your own, make sure you have a good site for it. The area needs to be fairly exposed, well away from trees and buildings that may obscure the view, and so most astronomers site their observatories on the top of a hill. You may not have such an opportunity, and you may have to put your observatory in your garden or a nearby field, but you should still watch for anything that may potentially block your view. You'll want to make sure that there aren't any bright outside lights nearby too, because these can dazzle you and spoil the view.

Setting Up Your Telescope

Getting your first telescope is an exciting event; you'll probably want to take your telescope outside on the first night and start using it straight away. You'll be able to view a lot of amazing sights in the night sky that you simply couldn't see without a telescope.

But using a telescope for the first time can be quite daunting. Unless you follow a few simple steps, you may end up getting frustrated and losing patience with your new telescope.

Telescopes have a lot of different components and moving parts, and you should try and become familiar with all of these before you try and use your telescope for stargazing (see Figure 4-4).

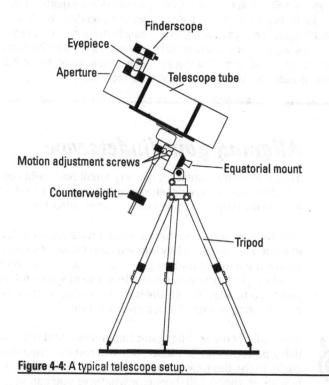

Finderscope

Eyepiece

Aperture

Telescope tube

Motion adjustment screws

Equatorial mount

Counterweight

Tripod

Figure 4-4: A typical telescope setup.

Before you step outside

Setting up your telescope indoors first is always a good idea so that you can see what you're doing. Setting it up outside at night, whether it's the first or tenth time, is very tricky, so you'll want to get the hang of it when you can see what you're doing.

Make sure that you have all the parts and that they're all working properly. You may want to find a small box to store all the loose parts in – the eyepieces, the front cover, the lens caps and so on.

Ask the experts

If you're still not sure which telescope to buy or if you're having trouble operating it, you should think about joining your local astronomical society. Hundreds of such societies exist around the world, and their members are experts on stargazing kit and operation. You'll get a lot of help from them, so much so that one day maybe you'll be the astronomer that people come to for help with their telescopes!

Aligning your finderscope

Your telescope – unless it's a very small one – will come with a *finderscope*, a smaller telescope that attaches to the tube of your main scope and points in the same direction.

This finderscope has a much smaller magnification than your main scope, and so it has a much larger field of view, which makes it useful for finding things in the sky when you need a wide view. For you to get the best use of your finderscope, you need to align it with the main telescope so that they're both looking in *exactly* the same direction.

You can, of course, align your finderscope and telescope using a bright star or planet, but you can do it much more easily in the daytime using some distant landmark – a chimney, mast, pylon, hill, house or whatever you can see.

Aligning the finderscope in the daytime has many benefits:

- ✔ You can see what you're doing in daylight.
- ✔ Your target won't move, unlike the stars that appear to move as the Earth spins.
- ✔ You'll know for certain that you're looking at the same target, unlike at night when all stars look alike!

You may be surprised that your telescope and finderscope make things appear upside down; the image is inverted. This doesn't matter for astronomy (an upside down planet looks much the same as a right-way-up planet!) but may be confusing when you're looking at things down on Earth when aligning your finderscope.

You should *never* point your telescope at the Sun. If your alignment target is on the horizon with a clear sky behind it, make sure the Sun is nowhere nearby.

To align your finderscope and telescope:

1. **Get your alignment target in the centre of the field of view of your main scope.**

   Your scope may have come with a cross-hair attachment that you can fit to your eyepiece to help you get the target exactly in the centre.

2. **Adjust your finderscope using the small screws attached to it, so that your target is exactly in the centre of your crosshairs.**

3. **Keep rechecking and adjusting until you're sure your finderscope and telescope are looking in the same direction.**

   Your telescope and finderscope will be aligned. This means that when you're stargazing at night, you can use your finderscope to track things down and be confident that your target will be in view in your main scope, too.

After your finderscope and telescope are aligned, you need to make sure that you don't knock them back out of alignment again. Otherwise, you'll have to start from scratch. Don't ever use the finderscope as a handle to carry your telescope!

Focusing your telescope

Another useful daytime activity is learning how to adjust the focus of your telescope. Everyone's eyes are different, and you need to constantly re-adjust the focus of your telescope when stargazing with it. You'll want to know how to focus your telescope before you try it in the dark.

The eyepieces are what magnify and focus the light for your eyes. By slightly adjusting their position, you can change the focus so that blurry stars become sharp points of light. You adjust the focus near the eyepiece tube, usually by turning a small knob or a dial. (All telescopes are a little different, so you should look in your telescope instruction manual to help you find your focus control.)

After you find the focus control, point your telescope at a distant target (not anywhere near the Sun!) and adjust the focus so that you get a sharp image. Practise changing the focus while looking through the eyepiece, finding the focus control and adjusting it without looking. After you get the hang of focusing, you should be able to adjust the focus at night to make all the stars appear as perfectly sharp dots.

Telescopes are built to look at very distant objects, such as stars and planets, and they won't be able to focus very easily on nearby targets. Don't be surprised if you can't bring a nearby target into perfect focus; try looking for something a bit farther away.

If you ever look through your telescope and see faint rings of light instead of sharp dots, don't panic! These rings are still stars, but your telescope is so out of focus that they're blurred out. Try making large adjustments to your focus so that these rings get smaller and smaller before focusing as perfect individual dots.

Setting up and cooling down

After you have set up your telescope indoors and become familiar with it, then you can take it outside and do some stargazing. Stand your telescope on a safe flat area, away from any potential hazards that you may fall into in the dark, such as holes or ditches. You should try to position your telescope so that you have a large area of clear sky overhead (so well away from trees or buildings). You'll also want to make sure that no bright lights are nearby, because they can spoil your night vision and dazzle you as you're stargazing.

To get the best out of your telescope, you should put it outside an hour or so before you're about to use it, to give it a chance to cool down (assuming that the weather is cool). If you take your telescope from a warm room indoors to cooler temperatures outside, then the telescope metal and the air inside the telescope tube will be hotter than the cooler night air. This difference in temperature can make the air move inside the telescope, giving you a wobbly, blurry image. Make sure that all the covers and lens caps are still on your telescope, to prevent moisture forming on the important bits. If, when you take your telescope outdoors in the evening, you're not sure what the weather has in store, let your telescope cool down under some shelter, just in case it rains!

Aligning your equatorial scope

Stargazers with an alt-az or Dobsonian mount (see the earlier section on mounts) can start stargazing when they've set up their telescopes as described above, but if you have an equatorial mount, one final alignment will help you get the best out of your telescope: aligning the mount with the sky.

Somewhere on your equatorial mount is a scale with a knob, lock or dial that lets you adjust the mount for your latitude. You should adjust it so that the reading on this scale is the same as your latitude on Earth. You can find your latitude by using an atlas or map or by going online.

After you set your latitude correctly, one axis of your equatorial mount will be pointing up into the sky at an angle; this line is called the *polar axis*. You should now physically turn your entire telescope base so that this polar axis points towards your nearest pole (north in the northern hemisphere and south in the southern hemisphere). You can use a compass to get this just right, or if you're in the northern hemisphere, you can use the North Star (see Chapter 10).

After you make both of these adjustments, your polar axis will be pointing *exactly* at the North or South Pole of the sky, and you'll be able to track your targets by adjusting only one axis of your mount, the RA axis. (See the section on equatorial mounts above for more on RA.)

Moving about the sky

If your telescope is mounted on an alt-az or equatorial mount, knobs or dials will allow you to move the telescope around and find different targets in the sky or track the same target as the Earth spins.

Often, two knobs lock the telescope to prevent it swinging about too much. If you loosen these knobs one at a time, you should notice that one lock stops the telescope moving up and down, while the other stops it moving from side to side. By unlocking both at once, you can move your telescope around by hand through very large movements. Use this method to get the telescope pointing in roughly the right direction and then lock both knobs off again.

You then need to find the two dials that allow you to make fine adjustments about the axes after they're locked in place. These fine-adjustment dials will let you move the telescope gradually when you're looking through the eyepiece or the finderscope. You have to get used to finding the dials in the dark, though, without looking, which can take a lot of practice.

Getting Your First Look

After all the setup, you're ready to take your first look through your telescope (okay, so maybe you sneaked a peek already while you were setting it up . . .) and start your career as a serious stargazer.

On your first night out, make sure that you pick targets that will look great through your telescope – a thin crescent Moon or half Moon, Jupiter or Saturn, or one of the more spectacular faint fuzzies. You can begin with the binocular list in Chapter 3 and see how the objects in that list compare when viewed through a telescope.

The faint fuzzies up closer

After you've spent a few nights getting to know your telescope, you'll be more confident in finding those more elusive – and more rewarding – telescope treasures.

After you catch the stargazing bug, you'll want to chase down even the very faintest of faint fuzzies, which may appear tiny and indistinct, at the limit of your vision, even through your telescope.

In addition to the items described in the following sections, a wealth of fainter fuzzies are worth tracking down using your telescope, far more than your eyes or binoculars will see. I don't list them all here because there are so many, so go to Part III for a constellation-by-constellation list of good telescope targets.

Gas clouds

The **Orion Nebula, M42** (see Chapter 11), is a stunning telescope object, even when you are using small telescope. You can easily see it with the naked eye from a dark site, but if you look at it through a telescope, you'll be able to see its structure: a grey, nebulous cloud of gas in space. If your skies are dark, then you may make out some gaseous arms extending out from the main bulk of M42. You should also see the four brightest stars inside the Orion Nebula, known as the *Trapezium,* shining like tiny jewels inside the cloud.

The **Eta Carinae Nebula, NGC 3372** (see Chapter 12), is bigger and brighter than the Orion Nebula, but because it's visible only in the southern hemisphere, it's not nearly as well known. Through a telescope, you see an oval shape – a large fuzzy cloud of gas in the sky – and if you're observing under dark skies you should make out some structure within the fuzziness, in particular a distinct L-shaped dark band running through it.

Galaxies

The **Andromeda Galaxy, M31** (see Chapter 14), is the closest big galaxy to the Milky Way. On a dark night, you can see the smudge of the Andromeda Galaxy with the naked eye, but a telescope will make it much larger. The Andromeda Galaxy has spiral arms, just like the Milky Way, and through a large telescope you may be able to see them. Medium-sized telescopes probably won't show the spiral arms, but you may be able to see the two fainter galaxies that orbit around the larger Andromeda Galaxy – M32 and M110.

The **Large and Small Magellanic Clouds** (see Chapter 15), are companion galaxies to the Milky Way and appear as very large faint fuzzies to the naked eye. But turn even a small telescope on them, and you'll see that they are full of faint fuzzies themselves – star clusters and nebula inside the Magellanic Clouds. The Magellanic Clouds are southern hemisphere objects only.

Open clusters

The **Pleiades, M45,** or the *Seven Sisters* (see Chapter 11), can actually look better through binoculars than through a telescope, but a telescope will let you see even more stars in this beautiful open cluster.

To get the best view, you'll need to use a low-power eyepiece, giving as big a field of view as possible.

The **Southern Pleiades, IC 2602** (see Chapter 12), is an open cluster just like the Pleiades, but it's a bit smaller and not quite as bright. Even still, it will more than fill the field of view of most telescopes. It's in the southern hemisphere, though, and so won't be visible to northern stargazers.

The **Double Cluster in Perseus, NGC 869 and 884** (see Chapter 14), looks beautiful through a small telescope. The two star clusters side by side make it unmistakable, and it's easy to find with the naked eye, too, lying half way between the bright constellations Cassiopeia and Perseus. A more powerful telescope with a smaller field of view will let you zoom in on these clusters one at a time.

Globular clusters

Omega Centauri, NGC 5139 (see Chapter 12), is the best globular cluster in the sky, although it's not visible to northern hemisphere stargazers. Through a telescope, you're able to see some of the individual stars that make up this object, whereas with binoculars, it just looks like a faint fuzzy blob.

47 Tucanae, NGC 104 (see Chapter 15), is another good southern hemisphere globular cluster. It's not quite as big and bright as Omega Centauri, so you'll need a medium or large telescope to make out the individual stars in this cluster.

The **Great Hercules Cluster, M13** (see Chapter 13), is one of the best globular clusters in the northern skies. If you have good dark skies and a medium-sized telescope, you should be able to see its individual stars.

Planetary nebulae

The Ring Nebula, M57, is one of the best planetary nebulae in the sky. Despite the name, a planetary nebula is nothing to do with a planet; it's a cloud of gas puffed off by a dead or dying star. They're called planetary nebulae because, through a telescope, they look a bit like planets. The Ring Nebula is bright enough to be seen with even a small telescope from the middle of a city, looking like a tiny oval, and larger telescopes will show its ring shape well.

The Dumbbell Nebula, M27, is another good example of a planetary nebula. A medium or large telescope should show you its dumbbell shape.

Double stars

Mizar and **Alcor**, in the constellation of Ursa Major (see Chapter 10), are among the best double stars in the sky. With just your eye, you may be able to tell that more than one star is here, but using a telescope, you'll see them really clearly. You'll also notice that the brighter of this pair, *Mizar,* is itself a double star, with a fainter companion very near the bright star.

Albireo is double star in the constellation of Cygnus the Swan (see Chapter 13), which you can easily separate out into its two components. Using a small telescope, you can see that these two stars are different colours. The fainter of the pair is blue, and the brighter is orange.

The planets through a telescope

The planets in the solar system are really where your telescope comes into its own. Even with a very good pair of binoculars on a tripod, the planets never get magnified much more than 20×, whereas even a small telescope can magnify them many more times.

With a telescope, you'll see the planets as definite disks, in most cases large enough to see features on. The longer you spend observing planets, the more you'll see, and it becomes very rewarding to track down the phases of Mercury, the polar caps on Mars, the cloud bands and Great Red Spot on Jupiter, and the gap between Saturn's rings.

Your high-power eyepieces will make the planets look much bigger, but they'll look dimmer overall. When it comes to a telescope's magnifying power, less is often more. If you can't see the details you're looking for when zoomed right in, put in a lower-power eyepiece and zoom out a bit; you'll be pleasantly surprised by how much more detail you may be able to see, even if the planet is much smaller than it was at high magnification.

Mercury

Through a large telescope, you should be able to spot that Mercury has phases, just like the Moon and Venus do. You'll need a much larger telescope – and perfect viewing conditions – to see any detail on Mercury's surface, but if you do succeed, you'll see darker blotches on the bright surface. These are areas on Mercury where the rock is a bit less reflective.

You should wait until the Sun has completely set before trying to find Mercury with your telescope. And to see Mercury at its best, you need to catch it at *maximum elongation,* when it's at its farthest from the Sun as seen from Earth, either after sunset or before sunrise.

Venus

A medium telescope may let you see the phases of Venus, but that's about all. Venus is shrouded in a thick atmosphere of carbon dioxide gas with lots of bright clouds floating and masking anything lying beneath. If you've got a larger telescope, you may make out different darker and brighter streaks in Venus's cloud tops.

Mars

The Red Planet, Mars, is a small planet – smaller than the Earth and Venus – but even a small telescope will let you see its orange disk. If you want to chase down some of the surface features of Mars, you'll need a medium or large telescope, through which you may see the darker regions on Mars, or even giant planet-wide dust storms that rage across the surface. If it's the right season on Mars, you may be able to see an icecap – a small white region at the top or bottom of the planet.

Jupiter

Jupiter is – for me – the best telescope planet, hands down. Through even a small telescope, you may be able to see cloud bands on the surface of the planet, as well as the *Great Red Spot,* a storm that's been raging in Jupiter's atmosphere for centuries. A medium or better telescope will let you see even more detail in Jupiter's atmosphere, and because the planet spins so often – once every ten hours – you can watch the cloudscape change as it spins.

Lining up with Jupiter are four of its largest moons: Io, Europa, Ganymede and Callisto. These four specks of light move around the planet at different speeds, with Io moving most quickly and Callisto most slowly. Over a few hours of a stargazing evening, you'll see these moons, known collectively as the *Galilean Satellites,* change positions, sometimes disappearing behind the disk of Jupiter, and at other times passing in front of the planet and casting a tiny dark shadow in its atmosphere.

The chances of anything coming from Mars . . .

Astronomer Percival Lowell spent many years observing Mars through a giant research-grade telescope in Flagstaff, Arizona. He convinced himself that the detailed and intricate dark markings on the surface of the Red Planet were not natural phenomena but constructions, built by Martians! Astronomers now know that they're just darker streaks on the lighter orange surface, and that if Mars ever did – or still does – have life, that life will be very primitive and most likely confined to underground.

Saturn

The mighty ringed planet Saturn is a firm favourite of telescope stargazers. You'll see the rings of Saturn clearly through even a small telescope, and the larger telescope you use, the more prominent they'll look. With large telescopes, you should be able to make out that the rings aren't continuous over their whole diameter; a gap, called the *Cassini Division,* appears in the rings, separating the inner ring from the outer ring.

Unfortunately, Saturn's rings don't always show themselves. Once every 15 years or so, you end up looking at the rings edge on, so you can't see them at all. The next time Saturn's rings will be edge on is in 2025. But between now and then you'll be treated to spectacular views, especially in 2017 when they're at their maximum tilt towards us.

Uranus and Neptune

The outer edges of the solar system are home to Uranus and Neptune, the Ice Giants. Through a medium telescope, you can see them looking like tiny bluish disks, with Uranus a lighter colour than Neptune and slightly larger, but neither has any features on its surface that makes it worth studying in detail.

Dwarf planets and asteroids

Some of the dwarf planets (Pluto, Makemake and Ceres) and the brighter asteroids (Pallas, Vesta and Juno) may be visible through very large amateur astronomy telescopes, but they show up only as tiny specks against the background stars.

Comets

Although bright naked-eye comets are rare, you can use a telescope to see plenty of fainter ones. Comets come and go as they orbit the Sun, so you should look up online what telescope comets are visible on the night when you're observing.

Chapter 5

Taking It Further: Astrophotography

....

In This Chapter

▶ Selecting a camera and other hardware

▶ Setting up your equipment

▶ Photographing your first image

....

*O*bserving the sky through binoculars or a telescope is hugely rewarding, but imagine if you could capture stunning images of what you see to share with the world. Astrophotography awaits, and with recent developments in equipment, amateur stargazers are beginning to obtain images that rival those that were possible with professional telescopes only decades ago.

Choosing the Right Camera

Most stargazers are content with just using their eyes, whether unaided or looking through a pair of binoculars or a telescope. You may be one of them, but every so often, you may see such a beautiful sight in the night sky that you want to photograph it.

Choosing the right camera is often difficult, but three basic camera types allow you to do something a little different:

✔ Single-lens reflex (SLR) cameras

✔ Point-and-click cameras

✔ Webcams and CCDs (charge-coupled devices)

SLRs

SLR cameras give you full control of the various manual settings, which is really important when you're doing astrophotography. They also let you see through the viewfinder *exactly* what the camera is seeing, so that you get a much better idea of what kind of image you'll get (unless you're taking a long-exposure shot, in which case the camera captures much more than your eye can see).

Old SLRs use 35mm film, but nowadays digital SLRs (known as DSLRs) are far more common. Their price is coming down so much that you don't have to break the bank to get one.

You need to have control over four basic settings if you want to get a good astro-image:

- ✔ Shutter speed
- ✔ Aperture size
- ✔ Sensitivity
- ✔ Manual focus

If you take astro-images, you do it at night, and your camera may automatically use the flash in such dim conditions. Turn off the flash! The flash will light up the foreground of your image and won't help you see any more stars.

Shutter speed

Most SLRs give you the option for controlling the shutter speed, using the *S* setting on your camera. You can vary the shutter speed from very slow (maybe as slow as 30 seconds) to very fast (maybe as fast as $\frac{1}{6000}$ of a second). Your camera displays a range of numbers. Those numbers with a ' after them are whole seconds; those without a ' are fractions of a second. For example, if you set your shutter speed to 2', your exposure time will be 2 seconds. If you set it to 60, it will be $\frac{1}{60}$ of a second.

For astro-imaging, you usually want a slow shutter speed, which gives you a long exposure image and much more opportunity to collect more light and therefore get a more detailed, brighter image.

Sometimes you can use the slow S-settings on your camera to give you a shutter speed you're happy with, but if you want much longer exposures – say, a five-minute exposure – you need to use the B setting, which stands for bulb.

When you push the exposure button when using the bulb setting, you can keep the shutter open for as long as you like; the shutter stays open until you let go of the button again, allowing you to make exposures for minutes or even hours.

 You should take lots of pictures of every scene at a variety of shutter speeds until you find an image you're happy with. Keep track of the settings you use for each picture, to build up a database of settings for future images. Keeping a database may seem time-consuming at first, but it will pay off later when setting up your camera becomes second nature, not a process of trial and error.

Aperture

The *aperture* of the camera is the hole where the light comes in. The larger the aperture, the more light your camera collects, and the brighter and clearer objects are. SLR cameras let you manually set the aperture using the *A* setting.

Avoiding star trails

If you're just using the camera to take a scenic shot with a starry sky in the background, you need to watch out for *star trails*. As the Earth spins, the stars overhead appear to move across the sky. So if you take a long-exposure image of a starry sky, you'll see the stars on your picture spread out into short lines rather than tiny dots. The longer the exposure you use, the longer the star trails are. Some of the most spectacular night sky images are those with star trains forming large arcs overhead, taken with an exposure of hours.

But if you want an image of the night sky as it appears to you, with dots of light rather than streaks, you have to use a shorter shutter speed, no more than 10 to 20 seconds. Stars near the poles move over a shorter distance than those above the equator, and so make shorter trails.

Camera apertures are described with an f-number, known as the *f-stop*. The usual f-stops on a camera are f/1.4, f/2, f/2.8, f/4, f/5.6 and f/8.

As you move up the list, each f-stop closes the aperture slightly and halves the amount of light entering the camera. For imaging faint stars in the night sky, you want wide apertures – so a low f-number.

Sensitivity (ISO numbers)

One useful feature of most DSLR camera is that you can set the sensitivity of your camera yourself. The camera's sensitivity is given by the ISO number. The higher the ISO number, the more sensitive you camera is to faint light, and so the brighter your image is (but the more grainy an image appears). Usually you want a high ISO for capturing dim stars, but once again, it's trial and error; take lots of images with different sensitivities to see what works best for you.

Manual focus

Most modern SLRs come with an autofocus feature, which you need to turn off while imaging the sky. Autofocus tends to get confused in very low light levels, and you probably won't get an in-focus image.

You should manually set the focus to infinity, represented by an infinity sign – ∞. To set the focus, turn the focus ring on your camera lens until the ∞ sign lies above the focus mark.

You may find that some lenses don't give perfectly focused images, even if you very carefully set the focus manually to infinity. In this case, you need to take a few images, slightly adjusting the focus every time. After you find the perfect focus position for your lens, mark that position on the focus ring using a pencil, so that you can find it more easily next time.

Point-and-click cameras

While not everyone has an SLR camera, point-and-click cameras are much more readily available – and cheaper! The compromise you make if you use a point-and-click camera is the inability (usually) to manually set the shutter speed, aperture size, ISO sensitivity, and focus. This lack of control means that it is hard to get just the image you want. However, you can cheat and still get a rather good image. (See the section 'Afocal astrophotography', later in this chapter.)

Webcams and CCDs

Dedicated amateur astronomers use modified webcams and hi-tech devices called CCDs (charge-coupled devices) that use a digital chip to gather light and to image the night sky. You can fix these items in place on the eyepiece holder of your telescope and connect them to a laptop or computer indoors. This means that you can image from the warmth and comfort of your own home!

Even cheap webcams have chips in them that give good astronomical images. You need to buy a webcam with a relatively high number of pixels, and you mustn't be afraid to modify it to fit into your telescope eyepiece.

Old 35mm camera film canisters are just the right diameter to slot into most telescope eyepiece sockets. Simply cut off the end of one of these canisters, glue the other end onto your webcam, and voilà – your very own astronomy webcam.

If you image through the eyepiece of your telescope, you need to have a motorised mount set up correctly for tracking the stars. Otherwise, you'll end up with very blurry images (or no image at all).

Figuring Out What Other Hardware You Need

The only hardware that's absolutely essential for astrophotography is a camera, but to help you get a good image, you may want to use some of the following kit, too:

 ✔ **Camera tripod:** A tripod allows you to fix your camera to a solid base so that you don't have to hold it in your hand. If you take exposures much slower than $\frac{1}{60}$ second, then the image becomes blurry because your hands shake, no matter how still you think you're holding the camera.

✓ **Camera cable release or remote control:** A cable release or remote control allows you to take images without touching the camera, which also helps reduce camera shake and image blurring. Most cable releases or remote controls allow you to lock the shutter open, meaning that you can start the image and leave the camera for the duration of the exposure.

✓ **Range of lenses for your camera:** If you're hoping to catch very faint objects with your camera, then consider getting a hold of a very fast lens – one in which you can open the aperture as wide as possible, which allows your camera to collect more light. Fast lenses, such as f/1.2 or f/1.4, can be very expensive, but the images you get are amazing. Another useful lens is a fish-eye. These fish-eye lenses have huge fields of view, so they're great if you're hoping to get a wide-angled shot with a lot of stars in it.

✓ **Telescope with a camera-mounting bracket:** Many telescopes come with a mounting bracket (see Figure 5-1) that lets you piggyback your camera on the body or mount of the telescope, basically letting you use your telescope as a very expensive tripod!

✓ **Telescope with motorised star tracking drive:** If your telescope has a motorised tracking drive, then you can afford to make much longer exposures of objects in the sky, without getting star trails.

✓ **Telescope with eyepiece adaptor:** For serious astrophotography of the planets or of deep-sky objects like galaxies and nebulae, you have to attach your camera to the eyepiece socket of your telescope. You'll need to get a special adaptor that fixes onto your camera in place of your camera lens and then slots into the eyepiece socket in place of an eyepiece. You essentially convert your telescope into one very powerful camera lens. If you plan on deep-sky astrophotography, then you'll need long exposure times as well as a telescope that tracks the stars automatically.

✓ **Filters:** Astronomers often use filters to block out unwanted light. Filters that block out light pollution can enhance your image a great deal, but you may find other filters are better for the image you want to capture.

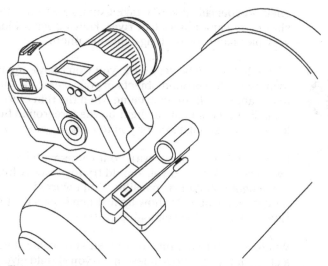

Figure 5-1: An SLR camera attached to a telescope with a mounting bracket.

Choosing Your Moment

After you assemble your astrophotography kit, you're ready to think about when you'll use it. If you're looking for a wide field of view for your image – say, you want to get a picture of the local scenery, but with a few bright stars or planets in the background – you may want to go out during twilight when the last light from the Sun can light up the area around you.

If you're intent on taking a long-exposure image or attaching your camera to the telescope eyepiece, then you should wait until the sky is as dark as possible, with no sunlight and no bright Moon up in the sky.

Next, you need to decide where you're going to take your image. Try to site your camera well away from bright local lights and far out of the glow of a town or city, if possible.

If you're looking for a scenery shot, then pick your place carefully. Long exposures cause any moving objects, such as branches of a tree blowing in the wind or passing people or cars, to become blurred. Sometimes, this motion adds to the effect – especially if you're taking an image near water, at the shore, or near a waterfall or river. Flowing water blurs in a very pleasant way in a long-exposure photo.

Given that it may be very dark, you won't have any lights to light up the foreground objects in your starscape photo. You may want to wait for an evening when the thin crescent Moon is up. Faint moonlight can light up the foreground but still leave a dark enough sky for you to capture your image.

If you're taking an image with your camera attached to the eyepiece socket, then you should try to get away from as much light pollution as possible. Don't worry if you can't, though; some objects show up just fine from a light polluted sky if you take a long enough exposure.

Whatever you're hunting for – whether a scenic starscape or a closeup through your telescope – you should give yourself the best chance possible to get a good image. Wait until your target is as high above the horizon as possible, above the murky air lower down towards the ground.

Astrophotography takes a lot of patience, so don't get frustrated if it doesn't work the first time you try it. By persevering, you can end up with some stunning images – not as good as those taken by the Hubble Space Telescope, of course, but even more important to you because you took them yourself!

Taking Your First Astro-Image

After you've got all of your kit assembled (see the section 'Figuring Out What Other Hardware You Need', earlier in this chapter), it's time to head outside and start snapping.

There are a few different ways you can capture astronomical images using a camera:

- ✔ **Starscape astrophotography** uses just the camera lens to capture the image, as you would with a daytime landscape shot. In starscape astrophotography, you can get images of entire constellations of stars at once.

- ✔ **Telescope astrophotography** uses your camera mounted onto the eyepiece socket of your telescope, which uses the telescope to collect much more light than your camera lens could, and so gives much more detailed close-up images of faint astronomical objects images. You need to buy adaptors to attach your camera to your telescope.

✔ **Afocal astrophotography** is much like telescope astrophotography, but rather than mount your camera onto the telescope eyepiece, you just point your camera down the eyepiece and snap a shot of what the telescope's looking at. This is much more straight forward, and requires far less effort, than proper telescope astrophotography, but you can still get some great images, especially of the Moon.

Starscape astrophotography

If you want to get a great starscape image of the night sky, you need to find a spot away from bright lights, where you can set up safely, and were you won't be disturbed by passing cars. Mount your camera on a tripod and a firm, flat surface, and set up your camera with these settings:

✔ **Shutter speed:** 10' (10 seconds)

✔ **Aperture:** f/3.5 (or as low as it goes)

✔ **ISO:** 1000 (or as high as it goes)

✔ **Focus:** infinity

✔ **Flash:** Off

✔ **Autofocus:** Off

Using your cable release, take a few photos using these settings and then start adjusting them one at a time to see what looks best:

1. **Increase the shutter speed to 20' or 30' or set it to *B* (bulb) and leave it open for much longer using the cable release or remote control lock.**

2. **After you find your ideal shutter speed, adjust the aperture up or down to see how the image changes as less and then more light gets in.**

3. **Try a few ISO settings, maybe as low as 400, all the way up to the maximum for your camera.**

The higher the ISO, the greater the number of faint stars that will appear, but the grainier your image will be.

Telescope astrophotography

Taking an image through your telescope with your camera fixed into the eyepiece socket is much trickier than taking starscape images (see preceding section), but essentially you go through the same process of adjusting the settings one by one until you find the right combination for you.

Make sure your telescope is set up correctly and is tracking the stars, so that your camera tracks too (see Chapter 4).

You may have to keep the shutter open for minutes at a time to get an image of that elusive faint fuzzy, so the time for each image is greater, meaning you get fewer images in the course of a night, and so fewer comparisons. Finding what's right for you requires many nights of imaging, but don't give up; the results will be worth it!

Afocal astrophotography

Afocal astrophotography sounds very hi-tech, but what it essentially means is holding your point-and-click camera up to the eyepiece of a telescope and taking an image of what your eye can see. Getting just the right image can be tricky, because you need to hold the camera still and in exactly the correct place to avoid parts of your image being blacked out.

The light leaving the eyepiece comes out in a narrow beam; your camera lens has to sit directly in that beam in order for you to get a good image.

Afocal astrophotography really only works well when you're imaging the Moon. You'll be able to see lots of detail on the Moon's surface in your image. For most other astronomical objects, afocal astrophotography won't work, because you need longer exposures to pick up all the light, and your hand-held camera will wobble too much.

Part II

Joining the Dots: Learning Your Way Around the Night Sky

The 5th Wave   By Rich Tennant

"...and that's the North Star. Knowing its location helps you chart a safe journey."

In this part...

The sky is full of amazing sights, from stars dotted in constellations, to planets and moons, comets and asteroids, galaxies and nebulae.

In this part, I guide you through each of these different types of object, helping you get to grips with what you're seeing overhead each night.

If you're wondering whether that bright point of light is a planet or a star, or whether that fuzzy patch you can see in your telescope is a galaxy or a gas cloud making new stars, this is the place for you.

Chapter 6

The Fixed Stars

*I*n this chapter, you find out about the most common of night sky objects – stars – as well as the many faint fuzzy objects that are associated with stars: giant galaxies, beautiful star clusters and distant clouds of star-forming gas.

These objects all stay in the same place in relation to one another. These fixed stars move across the sky as the Earth spins, some rising and setting, but they are fixed in patterns that won't change (actually they will, but not any time soon!).

Looking at a Night Sky Full of Stars

The night sky is full of stars. If you head out somewhere dark, away from the orange glow of light pollution, you can see almost too many stars to count. If you've got good eyesight and go to a truly dark-sky site, you may be able to see up to 4,000 to 5,000 stars. But the limit here is your eyesight, not the number of stars. In fact, there are between 200 *billion* and 400 billion stars in the Milky Way alone.

'A star is simpler than an insect'

Stars are basically giant balls of gas – mainly hydrogen, but other gases too. In the core of stars, the temperature and pressure are so high that the hydrogen begins to fuse together, in a process called nuclear fusion. In this process, four bits of hydrogen are turned into one bit of helium, plus a little bit of energy. This energy is what makes stars shine and forms the light that you see. The Astronomer Royal, Professor Martin Rees, summed it up nicely: 'A star,' he said, 'is simpler than an insect.' Larger stars can fuse helium together to make heavier elements, and then fuse those together to make others, and so on. Stars are the factories where all the elements heavier than helium are made.

Simply casting around the sky with a pair of binoculars enables you to see thousands more stars than your eyes can see alone. Do you see something that looks a bit fuzzy, not quite as sharp as the other stars? Chances are that fuzzy is something interesting that you may want to turn your binoculars or telescope towards.

Twinkle, twinkle, little star

Go outside tonight and look at the stars. See them twinkling? Thanks to nursery rhymes, most people have known from an early age that stars twinkle, but why do they?

The stars themselves aren't twinkling; it's the air in the atmosphere that makes stars twinkle. Starlight travels for hundreds of trillions of miles through the vacuum of space, only to end its journey in the last fraction of a second passing through the atmosphere overhead. If the air above you is turbulent, it'll bounce the light around, making the stars appear to twinkle. The air's more likely to be turbulent looking towards the horizon, because it rises from the warmer ground and cools, and so stars just above the horizon twinkle more than stars overhead.

Because stars twinkle most when you're looking at them low on the horizon, it means that to see a star – or a faint fuzzy, or a planet – looking its best, you should wait until it's as high above the horizon as possible.

Connecting the dots

While you're outside tonight and looking up, try to imagine joining the dots in the sky to make pictures, just as you would in a giant dot-to-dot book. You'll be doing what stargazers have done the world over for thousands of years – creating pictures in the sky, called *constellations*. The patterns that the stars make stay the same for thousands of years, so people have always seen the same patterns in the sky, even if they joined them up in different ways and gave them different names. (Part III offers a detailed guide to these pictures in the sky.)

The celestial sphere

Stars look like tiny sparkling dots embedded in a domed roof (see Figure 6-1), and for thousands of years, that's what people thought they were. This imaginary domed roof is called the *celestial sphere*, but even though astronomers know there isn't really a roof there, it's still a useful description of what the night sky looks like.

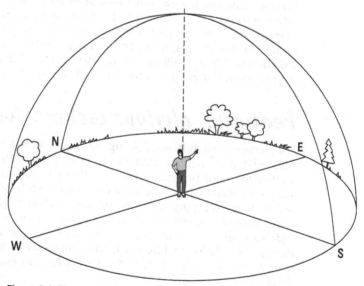

Figure 6-1: The celestial sphere.

Navigating the celestial sphere

Astronomers use coordinate systems to find objects the sky, just like the latitude and longitude coordinates used on Earth. Even though the sky isn't a celestial sphere surrounding the Earth, it's still useful to pretend it is for the purposes of navigating your way around the sky.

The coordinate system used most often by astronomers involves coordinates called right ascension (RA) and declination (dec). These coordinates are roughly the same as longitude and latitude on Earth. Finding a star's RA and dec in an atlas or list of stars will let you find it in the sky. The RA-dec coordinate system is an absolute system, in that it is fixed with respect to the stars. This means that as the Earth spins and the stars appear to move across the sky, so too does the RA-dec coordinate grid. Other coordinate systems, such as the alt-az (altitude and azimuth) system, are in use, but most astronomers use RA and dec. Alt-az is not an absolute system, in that it is based on your viewing location and time, and so a star's alt and az coordinates change as it rises and sets.

Now astronomers know that stars are, in fact, all different distances away, and the ones that look like they're next to each other in the sky may actually be huge distances apart, their closeness just being a line-of-sight effect. For example, if you hold your thumb up at arm's length and position it in front of a distant building or hill or tree, the two appear side by side in your view, but you know they're really far apart.

Featuring glorious technicolour

Stars come in all different colours, but you may not have noticed it. Most people think that all stars shine with the same white light. Look tonight and see whether you can spot any stars that aren't white. The most obvious off-white ones are *red giant* or *red supergiant* stars, old stars that are dying and, in the process, cooling down and changing colour. Stars like *Aldebaran* in Taurus, *Antares* in Scorpius, *Arcturus* in Bootes, *Betelgeuse* in Orion and *Gacrux* in the Southern Cross are all stars that look obviously red. (See Part III for star maps of these constellations.)

The fact that stars look more or less all the same colour boils down to your eyes' inability to make out colours in faint light rather than a lack of colour in the stars themselves. Binoculars and telescopes can help you see star colours, and some of the most beautiful sights in astronomy are fields of stars in glorious technicolour, looking like multicoloured jewels scattered on black velvet.

Oh be a fine girl/guy kiss me

Astronomers classify stars according to their temperatures using a series of letters: O, B, A, F, G, K and M, where O types are the hottest blue-white stars and M types are the coolest red stars. This series of letters is easiest to remember using the mnemonic 'Oh, be a fine girl/guy kiss me'! Earth's star, the Sun, is a G-type star, a middle-temperature yellow star. It's still pretty hot though, at around 6,000°C.

Here's what the letters mean:

O Blue (hottest, over 30,000°C)

B Blue-white

A White

F Yellow-white

G Yellow

K Orange

M Red (coolest, around 3000°C)

These letters don't seem to match the colours, but they exist for historical reasons too long and complicated to go into here!

While red stars might look hotter to you (red's a warm colour, after all)

they're actually the coolest of all. Just as when you heat a piece of metal it glows red hot then orange, then yellow, then white, so too with stars. The hotter a star is, the more blue it appears.

Stars are further classified by a number after the letter, from 0 to 9, which tells you whether that star is a cooler (0) or hotter (9) star for its type. Our star, the Sun, is classed as a G2, which means it's at the hotter (2) end of the G-type temperature range.

That letter–number combination in turn is followed by a Roman numeral telling you what type of star it is:

I Supergiants (the largest stars known)

II Bright giants

III Normal giants

IV Sub-giants

V Main sequence (medium stars like our Sun)

VI Subdwarfs (the smallest stars known)

(continued)

(continued)

These types of stars denote stars at different stages of their life cycle.

In Part III, the brighter stars in each constellation are listed with their spectral type. The Sun, for example, is a G2V star – that is, a relatively hot (2) yellow (G) main sequence dwarf (V) star.

There are some other lower-case letters added after the Roman numeral to further define what type of star it is, but that's beyond the scope of this book.

The Milky Way, the Sun's Local Galaxy

Another feature of a dark night sky that stays fixed in relation to the fixed stars is that band of grey light that stretches all the way across the sky, visible from anywhere on Earth if you're far from bad light pollution. This grey band is your home in the universe, a galaxy of up to 400 billion stars known as the *Milky Way*.

To the naked eye, the Milky Way looks like a haze of light, but turn your binoculars or telescope on it and wow! – you can see that it's full of stars. The light you see is the light from billions of stars, too far away for your eyes to make them out individually, but so abundant that you can still see their combined light.

The Milky Way is hard to spot in bright urban or suburban skies, but it's really obvious in dark sites. In some of the darkest places on Earth, the Milky Way can even cast a shadow.

Next time you're stargazing somewhere dark and see the Milky Way, look out for the dark lanes and blotches in it that appear to have no light. These blotches are clouds of dust in space within the Milky Way galaxy, blocking out light from the stars behind them, and seen in silhouette against the glow of the background stars.

If you look towards the centre of the Milky Way, you'll see many more stars and a lot of faint fuzzies, too. In this direction, the band of light that makes up the Milky Way in the sky is thickest and brightest.

Why does the Milky Way look like a band?

One of the most common questions asked by stargazers who see the Milky Way for the first time is 'Why does it look like a band?' The Milky Way looks like a band in the sky, because you're inside it. In reality, your galaxy is shaped a bit like a fried egg: the yolk is the galactic centre, and the white is the rest of the galaxy, orbiting the centre in a flat disk containing spiral arms. The Sun and Earth inhabit the disk of the galaxy, about two-thirds of the way out from the centre. From your position inside it, you can't see the whole galaxy; instead, you see it edge-on, as shown in the figure, and from that position, it looks like a band in the sky.

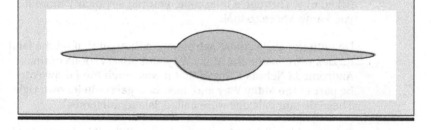

The galactic centre lies in the same direction as the constellation Sagittarius, and so the Milky Way looks best near this constellation. Unluckily for northern stargazers, Sagittarius and the galactic centre are highest in the sky and best seen from southern latitudes during their winter.

Getting Familiar with the Faint Fuzzies

Go out tonight and look around. What do you see? Stars, yes, but anything else? What's that faint fuzzy patch in the sky? Well, it may be an entire galaxy of stars, a cloud of gas lit up by the newborn stars inside it, a sphere of old stars orbiting the Milky Way or the remnants of a dead star. These objects are all referred to by astronomers as *nebulae*, meaning 'clouds', but are also known as *'faint fuzzies'* because they're, well, faint and fuzzy.

Faint fuzzies come in several forms:

- ✔ Galaxies
- ✔ Globular clusters
- ✔ Open clusters
- ✔ Newborn stars
- ✔ Dead stars

Galaxies

Beyond the Milky Way is a vast universe full of other galaxies, and some of the closest galaxies to you are visible to the naked eye. Through a telescope, you can see plenty more if you know where to look.

Up until the 1920s, most astronomers thought that all the faint fuzzies were inside the Milky Way. But observations of the Andromeda Nebula showed that it was much too far away to be part of the Milky Way and must be a galaxy in its own right. These distant galaxies were called 'island universes'.

Galaxies, like all faint fuzzies, are quite difficult to locate, but you can find many of them listed in the constellation guides in Part III and marked on the maps there, too. Some of the brightest galaxies are:

- ✔ Large and Small Magellanic Clouds (see Dorado and Tucana, both in Chapter 15)
- ✔ The Andromeda Galaxy (see Andromeda in Chapter 14)
- ✔ The Centaurus A Galaxy (see Centaurus in Chapter 12)
- ✔ The Sombrero Galaxy (see Virgo in Chapter 12)
- ✔ The Whirlpool Galaxy (see Canes Venatici in Chapter 12)
- ✔ The Southern Pinwheel Galaxy (see Hydra in Chapter 12)

Globular clusters

Another great sight through binoculars or a telescope is the round fuzzy globular clusters, each one made up of hundreds of thousands of old stars in a sphere. Around 150 globular clusters are known, and they all sit in a halo around the Milky Way galaxy, well outside the disk and central bulge.

Calculating distance

The distances to far-flung galaxies are calculated using what are called *standard candles* – objects inside those galaxies that astronomers know the exact brightness of. (As an example, see δ Cephei in Chapter 10.) Astronomers also use the *distance-squared law*, or *inverse square law*, which says that the amount of light you receive from an object drops off in inverse proportion to the distance squared. So if you get twice as far away from a star, it will appear four times fainter. Using standard candles and the distance-squared law, you can tell how far away things are.

 The stars in globular clusters are probably some of the first ones formed in the galaxy, which means that they're very old stars, more than 10 billion years old.

 You can see only a handful of these fuzzy globular clusters with just your eyes, but binoculars or a telescope let you see more of them, hanging like celestial baubles in space.

Some of the brightest globular clusters are:

- ✔ Omega Centauri (see Centaurus in Chapter 12)
- ✔ 47 Tucanae (see Tucana in Chapter 15)
- ✔ The Great Globular Cluster (see Hercules in Chapter 13)

Open clusters

As well as galaxies and globular clusters (see preceding sections), you can find lots of faint fuzzy *open clusters*. Open clusters are collections of a few hundred or a few thousand stars inside the galaxy.

All the stars in an open cluster were born in the same cloud of gas, but those stars haven't yet moved away from where they were formed. You can think of them as teenage stars, all still hanging about together but getting ready to move away, settle down and raise a family of planets.

Stars in open clusters are very young, at most only a few hundred million years old. That sounds pretty old for teenagers, but compared with the 4.5-billion-year age of the middle-aged Sun, it's still pretty youthful. Open clusters like the Pleiades contain hot massive stars less than 100 million years old.

Open clusters look great through binoculars or telescopes. Some of the very best ones to look for are:

- The Pleiades, or Seven Sisters (see Taurus in Chapter 11)
- The Beehive Cluster (see Cancer in Chapter 12)
- The Southern Pleiades (see Carina in Chapter 12)
- The Jewel Box (see Crux in Chapter 15)
- The Double Cluster (see Perseus in Chapter 14)
- NGC 2477 (see Puppis in Chapter 11)

Newborn stars

Stars begin life inside clouds of hydrogen gas floating about in the galaxy. As stars are born and begin shining, they can light their cloud up from the inside, letting you see these stellar nurseries.

Stars form out of clouds of hydrogen gas – the gas that was formed in the Big Bang – but that hydrogen has been mixed and enriched with other gases, too – heavier elements that were blasted off dying stars in the last moments of their lives. It's these heavier elements that form planets like the Earth.

The brightest and most dramatic of all these gas clouds is the Orion Nebula, which is actually just the brightest bit of a much bigger cloud called the Orion Complex, which contains other gas clouds like the Horsehead Nebula.

Some gas clouds worth looking for with binoculars or telescopes are:

- The Orion Nebula (see Orion in Chapter 11)
- The Eagle Nebula (see Serpens in Chapter 13)

✔ The Triffid Nebula (see Sagittarius in Chapter 13)

✔ The Lagoon Nebula (see Sagittarius in Chapter 13)

Dead stars

After you've had your fill of looking at newborn stars, you can turn your attention to the dead ones! When massive stars die, they explode, each blasting off its outer atmosphere in a giant explosion called a *supernova*. The gases that are blasted off are lit up by the energy of that explosion, and you can see the resulting faint fuzzies – called *supernova remnants* – using binoculars or telescopes.

Some of the best are:

✔ The Crab Nebula (see Taurus in Chapter 11)

✔ The Veil Nebula (see Cygnus in Chapter 13)

✔ The Vela Supernova Remnant (see Vela in Chapter 12)

Small stars undergo less dramatic deaths. Instead of blasting off their atmospheres, small stars puff them off in a final coughing fit, but they, too, can form faint fuzzies worth looking for. This type of fuzzy is called, rather misleadingly, a *planetary nebula*. They're called that because they do look a bit like planets when you see them through a pair of binoculars or a small telescope.

Some of the best planetary nebula faint fuzzies are:

✔ The Ring Nebula (see Lyra in Chapter 13)

✔ The Dumbbell Nebula (see Vulpecula in Chapter 13)

✔ The Cat's Eye Nebula (see Draco in Chapter 10)

✔ The Eskimo Nebula (see Gemini in Chapter 11)

✔ The Eight-burst Nebula (see Vela in Chapter 11)

✔ NGC 3918 (see Centaurus in Chapter 12)

Messier and Messier: Cataloguing the Faint Fuzzies

Astronomers list the faint fuzzies in a variety of catalogues. Most of the brighter ones have got proper names, but every so often, you'll stumble over one that's just listed as a number. The most famous of faint fuzzy catalogues is the list of *Messier objects,* named after French astronomer Charles Messier. Objects in the Messier Catalogue are given an M-number, and 110 faint fuzzies are in that list, including a mix of galaxies, globular clusters, open clusters, supernova remnants and planetary nebulae.

Other catalogues include: the longer New General Catalogue, in which 7,840 objects have been given NGC numbers; the Index Catalogue (IC); and the Caldwell Catalogue, created by British astronomer Patrick Moore as a list of 109 interesting faint fuzzy objects for stargazers to hunt for using their telescopes.

Tracking down faint fuzzies can be difficult, especially when you have to use telescopes or binoculars to find them. And they won't look anything like those amazing Hubble Space Telescope pictures you may have seen – in particular, you almost certainly won't see any colour in a faint fuzzy when looking through a telescope; they all appear greyish-white, because our eyes can't pick up colour in faint objects such as these. However, your eyes, when you're observing these objects, are absorbing particles of light – photons – that were created inside a distant nebula, which is an amazing thing to remind yourself of when you're gazing at a faint fuzzy.

Chapter 7

The Wanderers

- -

In This Chapter

▶ Observing the Sun and the Moon

▶ Identifying the naked-eye planets

▶ Finding the fainter wanderers

▶ Observing shooting stars

▶ Locating satellites

- -

*Y*ou can see some really exciting objects moving through the sky against the background of the fixed stars and the patterns of the constellations. In this chapter, you meet these wanderers.

Identifying the Wanderers

Chapter 6 talks about the fixed stars and faint fuzzies and how the stars are fixed into patterns called constellations. But sometimes you see different objects in amongst the familiar constellation shapes. In most cases, these objects are one of the five naked-eye planets, Mercury, Venus, Mars, Jupiter or Saturn, but the Moon and the Sun also move against the fixed stars, so they are called wanderers, too.

The brightest of the wanderers are visible to the naked eye, and stargazers have seen them in the sky for thousands of years. The ancient Greek astronomers called them *planetai*, which translates as 'wanderers', and is where the word 'planet' comes from.

Of course, you'll know when you're looking at the Sun or the Moon, but the other wanderers aren't quite so obvious;

in many cases, they look like very bright stars. This section helps you identify your Aries from Uranus and discover how to spot the planets.

In Chapter 1, you see how the sky changes over days, months and years. Most of these changes are down to the movement of the Earth as it spins about its axis and orbits around the Sun, or down to the movement of the Moon as it orbits around the Earth.

If you were in a spaceship hundreds of millions of miles above the Earth's North Pole and looking down, you'd see the Earth spinning once a day anticlockwise, like a top slightly tilted to one side, and going around the Sun once a year, orbiting in an anticlockwise direction, with the Moon going around the Earth once a month, also in an anticlockwise direction.

But you're not in a spaceship looking down from above; you're on the Earth, and so you see the Sun, Moon, planets and stars appear to rise and set once a day as the Earth spins. You'll also see the Sun, Moon and planets appear to move from day to day against the background stars.

The wandering Moon

The Moon is one of the wanderers, and it's very simple to see its motion from night to night. Try it for yourself:

1. **Next time the Moon's up in the night sky, note its location in relation to the stars around it.**

   You may even want to draw a picture of its position or mark it on a star map that you already have.

2. **Go out the next night and mark the Moon's new position.**

   The Moon will have moved! It will be a little bit farther east relative to the stars than it was the previous night, and its phase will have changed.

3. **Keep marking the Moon's position from night to night, and you'll begin to see the path that it follows against the background stars.**

   After about one month (specifically 29.5 days), you'll see that the Moon has got back to roughly where it started, and the whole cycle will repeat itself.

The Moon by degrees

How far does the Moon move in the sky from night to night? Try to work it out. The Moon makes one complete circuit against the fixed stars over 29.5 days. Round it up to 30 days, to make your calculations easier. To make one complete circuit around the sky, the Moon has to move through 360 degrees, which means it moves about 12 degrees each day, on ⅟₃₀ of a complete circle, against the fixed stars. Twelve degrees is about the same size as your clenched fist held at arm's length. This amount is quite a distance in the sky, and so the Moon may spend only a couple of nights in each constellation that it's passing through.

Figure 7-1 shows what your drawing of Moon phases might look like over the course of a week, as the Moon moves from a half Moon (first quarter) to a Full Moon, passing through Taurus and Gemini. Northern hemisphere stargazers see the Moon move from right to left from night to night; southern hemisphere stargazers see it move from right to left, but it always moves from west to east.

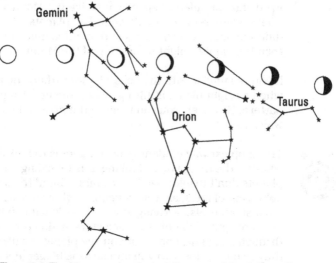

Figure 7-1: The Moon moving through Taurus and Gemini over the course of a week.

The wandering Sun

The Sun moves against the background stars, too, changing its position as the Earth orbits it once a year. The Sun moves much more slowly than the Moon against the fixed stars, making one complete circuit of the sky in a year rather than in a month as the Moon does.

And you can't draw the Sun's position against the fixed stars because you can't see those stars in the daytime (except during an eclipse!). But you can keep track of which stars are visible just as the Sun sets each day, and you'll notice that over the course of the year, different stars are visible at sunset in different seasons.

Don't twinkle, don't twinkle, little planet

As well as noticing the Sun and the Moon moving against the fixed stars, ancient astronomers also spotted some other things moving. These objects looked a bit like bright stars, but they weren't fixed in constellations; they moved against the fixed stars at different speeds, just like the Moon appears to. These wanderers are the planets. The planets all wander at different speeds, over weeks and months, much more slowly than the night-to-night wandering of the Moon.

If you're familiar with the constellations, then you may notice when a bright planet is sitting among the usual stars of that pattern. However, you can identify the planets in other ways, too.

The simplest way of identifying a planet is to look and see whether the thing you're looking at is twinkling. As a rule, planets don't twinkle, while the stars around them do. The light from planets, as well as being brighter than that from most stars, is also arriving at your eyes in a much thicker beam of light. You only ever see stars as tiny dots, even through a big telescope, but you see planets as disks, because they're much closer to you and so look bigger. It's much harder for the air above you to knock these thick beams of light around, and so planets are much steadier and rarely twinkle.

Following the Zodiac

Just as you can mark the Moon's position each night against the fixed stars and constellation patterns (see the section 'The wandering Moon', earlier in this chapter), you can mark the planets, too. The planets also make paths against the fixed stars.

If you mark the positions of all the wanderers over the course of a year, you'll begin to join up these paths into one line that circles the entire sky (see Figure 7-2). This line is known as the *zodiac*, and its more scientific name is the *ecliptic*. This ecliptic line passes through the signs of the zodiac, which comprises 12 familiar constellations:

- Aries
- Taurus
- Gemini
- Cancer
- Leo
- Virgo
- Libra
- Scorpius
- Sagittarius
- Capricornus
- Aquarius
- Pisces

Figure 7-3 shows the path of the zodiac through the constellations Taurus and Gemini.

Whenever you see a planet, the Sun or the Moon in the sky, it always lies somewhere in one of the signs of the zodiac.

 You can find out where a planet is on the ecliptic by looking it up online – a search for 'Jupiter position April' should give you lots of websites that will tell you where to look for Jupiter in April – but you may want to be better prepared and subscribe to an astronomy magazine. These monthly publications list all the planets that are up in the sky for the particular month, and feature maps that show you where to find them.

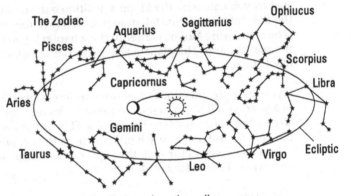

Figure 7-2: The constellations along the zodiac.

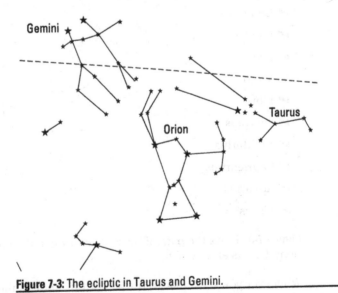

Figure 7-3: The ecliptic in Taurus and Gemini.

The thirteenth sign

Contrary to popular belief, the zodiac actually has 13 signs, not 12. The thirteenth sign is called Ophiuchus the Serpent-bearer (see Chapter 13). Don't worry if you've never heard of him – most people haven't! He sits between Scorpius and Sagittarius, and so the Sun, Moon and planets all spend a bit of time in Ophiuchus each time they move around the sky.

Shining Brightly: The Sun

The Sun is Earth's local star. It looks much bigger, brighter and hotter than the stars you see at night, but that's only because you're much closer to the Sun than you are to the other stars.

For example, say that you got into a spaceship – the fastest ever built – and flew towards the next-closest star, α Centauri. In a few *million years* after you arrive there, if you look out of your spaceship window, α Centauri will look just like the Sun does from Earth; if you look back at the Sun, it will appear as a speck of light in the black night sky.

Observing the Sun safely

Observing the Sun is incredibly dangerous. You should never look at the Sun directly, even with your naked eyes, and especially not with a telescope or binoculars. If you do, you can easily blind yourself.

So how can you observe the Sun safely? The best way is to join a local astronomical society or club and get the members to show you the Sun through their telescopes. So that they can see the Sun safely, these astronomers fit their telescopes with a *solar filter* that blocks out almost all the Sun's light, with only a tiny bit getting through.

Filters are best fitted at the front of a telescope; you can get filters that go onto the eyepiece, but they can often overheat and crack – and you don't want that to happen while you're looking through them!

Another good way of observing the Sun is by using a tele-scope to project an image onto a piece of paper or card, as shown in Figure 7-4.

If you point a telescope at the Sun, don't look through the tele-scope! You can hold a sheet of paper far from the eyepiece and see a round image of the Sun. This works using binoculars too.

You can buy specialist telescopes for observing the Sun safely that have built-in filters to block out most of the light. These are great for observing the Sun, but can't be used for anything else.

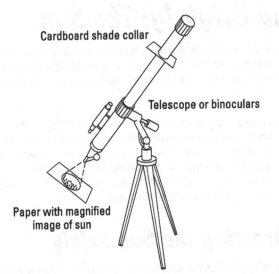

Cardboard shade collar

Telescope or binoculars

Paper with magnified image of sun

Figure 7-4: Projecting an image of the Sun onto a piece of paper.

Looking at sunspots

When you see the Sun through a filtered telescope or pro-jected onto a card, you may at first glance think it's just a boring, featureless object – but look more closely! Those tiny black specks are *sunspots*, cooler regions of the Sun's surface. They may not look like much, but even the smallest visible sunspot is bigger than the Earth.

Up and down: The solar cycle

You can't always see sunspots or solar flares on the Sun, because they only appear in great numbers when the Sun is very active. The Sun goes through a cycle of activity. Every 11 years or so, the Sun gets very active, before becoming more inactive. Then around 11 years later, the Sun becomes active again for a few years. When the Sun's very active, you get more sunspots and more flares.

If you're looking through a telescope that's safely fitted with a solar filter, look out to the edge of the Sun (the part that astronomers call the *limb*), and you may be able to see small loops (*prominences*), arcs or jets (*flares*) of gas. This gas has been blown off the surface of the Sun in an eruption caused by the Sun's magnetic field lines becoming tangled and then snapping.

Watching aurorae displays

If a particularly powerful eruption happens on the Sun, a large cloud of gas may blast from its surface. When this blast happens, the charged particles can fly through the solar system, and some of them even hit the Earth.

But don't panic; these particles are mostly harmless, and their main influence is seen high in the atmosphere where they make the air molecules glow. When this blast happens, stargazers at far northern or southern latitudes may observe the *aurora*. In the north, it's called the *aurora borealis,* or Northern Lights; in the south, it's known as the *aurora australis,* or Southern Lights.

At their most dramatic, aurorae can look like giant colourful curtains of fire in the sky. Their most common colour is green, but you may also see reds, blues or purples mixed in, too. And they move around in the sky, shimmering and flowing.

Displays of aurorae are some of the most beautiful natural phenomena and well worth making the effort to see, but astronomers can't predict them much more than a couple of days in advance, which makes planning for them tricky! If you do hear of a good display coming, then you should look north if you're in the northern hemisphere and south if you're in the southern hemisphere. A good clear horizon free of light pollution will help you see them.

Websites such as spaceweather.com let you keep an eye on the Sun's activity and can alert you when a good display of the aurora is due.

Different cultures have different names for aurorae. My personal favourite is the name given to them in Orkney in the north of Scotland, where they're known as the Merry Dancers!

Catching a glimpse of a solar eclipse

One of the most stunning of all natural phenomena is a total *solar eclipse.* An eclipse occurs when the Moon passes in front of the Sun, perfectly blocking out the disk of the Sun. When an eclipse happens, you see the Sun's beautiful outer atmosphere, called the *corona* or crown. The corona is there all the time; it's just that the Sun's light normally drowns it out.

That an eclipse happens at all is just a huge coincidence. The Moon just happens to be the right size and distance from the Earth so that it appears to be *exactly* the same size as the Sun in the sky. The Sun is *much* bigger than the Moon (its diameter is 400 times greater than the Moon's), but it's also *much* farther away (400 times farther from Earth than the Moon), and these factors conspire to make an eclipse possible.

If the Moon was a bit smaller or farther away from Earth, it wouldn't block out all the Sun's disk in an eclipse. If the Moon was a bit bigger or closer to Earth, then it would more than cover the Sun's disk, and you wouldn't see the corona nearly as well.

Eclipses are very rare from individual places on Earth, but a total solar eclipse occurs somewhere on Earth at least every few years.

In addition to total solar eclipses, you can also see partial eclipses, when the Moon only blocks out part of the Sun's disk. These aren't quite as impressive as total eclipses, but still well worth watching out for.

Eclipse chasing

Because eclipses are so spectacular, a tourism industry has been built around chasing them, with people traveling to sites where an eclipse is visible.

Eclipse chasing is great fun. For most people, seeing an eclipse is a once-in-a-lifetime experience. You need to plan well in advance: eclipses are so popular that hotels in the places where the will occur may start to book up years before! Many companies offer eclipse tourism packages and can help you maximise your chance of seeing an eclipse. But remember, you're at the mercy of the weather; all the planning in the world won't help you if it's cloudy!

Observing the Moon

The Moon is the Earth's nearest neighbour in space, and it can be the brightest thing in the night sky. It's also one of the most amazing sights seen through even a small telescope.

If you look at only one thing through a telescope, it should be the Moon. But you need to observe it in just the right conditions. The Moon at its brightest (during a full Moon) can actually be a hindrance to astronomy, because it drowns out the light of the fainter stars; it's like natural light pollution.

Phases of the Moon

The Moon goes through its cycle of phases once every 29.5 days. During a new Moon, when the Moon lies in the same direction as the Sun when seen from Earth, the Moon isn't visible in the skies at all. A few days after the new Moon, the Moon will start to appear as a thin crescent before growing to a half Moon and then a full Moon. The Moon then appears to shrink again on its way back to the new Moon.

The Moon's phases are:

- New
- Waxing crescent
- First quarter

- Waxing gibbous
- Full
- Waning gibbous
- Last quarter
- Waning crescent
- New

Figure 7-1 shows what the Moon looks like during each phase.

For you stargazers intent on chasing down the faintest stars or the faint fuzzies, anything brighter than a crescent Moon can hamper your chances, so you'll need to time your observation with a new Moon or, at the very most, a thin crescent Moon.

The Moon itself makes an amazing target for stargazers, especially when it's a thin crescent Moon. If you want to begin observing the Moon with a telescope or binoculars, you may want to buy a Moon map to guide you on your way.

Some of the obvious features you'll see on the Moon's surface when you look at it through binoculars or a telescope are shown in Figure 7-5, and are the:

- **Aristarchus crater** is the brightest crater, visible to the naked eye in the upper left portion of the Moon when seen from the northern hemisphere.

- **Copernicus crater** is a large bright crater with prominent bright rays, situated left of centre of the Moon when seen from the northern hemisphere.

- **Kepler crater** is another large bright crater, located left of Copernicus when seen from the northern hemisphere.

- **Tycho crater** is a large crater surrounded by numerous smaller craters, in the lower left portion of the Moon when seen from the northern hemisphere.

- **Mare Crisium,** the Sea of Crises, is a large, flat, dark plain in the upper right portion of the Moon when seen from the northern hemisphere. Although the Moon is covered in patches called *mare,* Latin for sea, there's no water there; the name *mare* comes from the days when people thought there might be oceans on the Moon. Now we know better; it's a dry lump of rock, and the *mare* are simply flat plains of dark rock.

Repeat after me: Half the Moon is always lit!

The Moon is a sphere of rock, and as the Sun shines light out into the solar system, half of the Moon's surface is always lit. It doesn't matter that you can't always see the entire lit half; the Moon is always half lit by the Sun. As the Moon orbits the Earth, sometimes that lit half is pointing right at you, so you see all of it (a full Moon), but sometimes the Moon's at such an angle that you see only part of the lit half. Maybe you see half of the lit half (a half Moon) or sometimes just a sliver of it (a crescent Moon), but half the Moon is always lit.

- ✔ **Mare Fecunditatis,** the Sea of Fecundity, is located just below and to the right of the famous Sea of Tranquillity (see below) when seen from the northern hemisphere.

- ✔ **Mare Imbrium,** the Sea of Rains, is the largest of the Moon's 'seas', located just above Copernicus crater when seen from the northern hemisphere.

- ✔ **Mare Nectaris**, the Sea of Nectar, lies just below the famous Sea of Tranquillity (see below) when seen from the northern hemisphere.

- ✔ **Mare Serenitatis,** the Sea of Serenity, lies just above the Sea of Tranquillity when seen from the northern hemisphere.

- ✔ **Mare Tranquillitatis,** the Sea of Tranquillity, is perhaps the most famous feature on the Moon, because it was the landing site for Apollo 11, when Neil Armstrong and Buzz Aldrin became the first men to set foot on the Moon's surface.

The Moon in Figure 7-5 is how it appears if you're stargazing in the northern hemisphere; southern hemisphere stargazers will see it upside down.

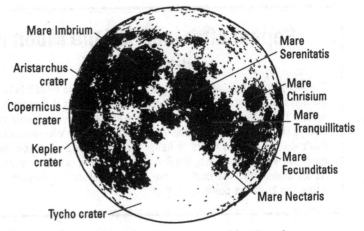

Mare Imbrium

Aristarchus crater

Copernicus crater

Kepler crater

Tycho crater

Mare Serenitatis

Mare Chrisium

Mare Tranquillitatis

Mare Fecunditatis

Mare Nectaris

Figure 7-5: Some of the obvious features on the Moon's surface.

The terminator

The best bit of the Moon to look at is the line between the lit part and the unlit part, which astronomers call the *terminator line*. If you were standing on the terminator line on the Moon, the Sun would be on the horizon, just rising or setting, and so the shadows are longest, just as on Earth at sunrise or sunset. Consequently, everything on the terminator – the mountains and the craters – stands out very dramatically.

And from night to night, as the Moon orbits the Earth, the terminator line moves across the Moon, showing different landscapes every night. Moongazers can therefore spend nights on end outside tracking down all the different features on the Moon's surface. When the Moon is nearly full, though, it can be quite dazzlingly bright if you look at it through a telescope. You can buy lunar filters that block out some of the Moon's light and make it more comfortable to observe when nearly full. A Moon atlas is a great purchase that can help you find your way around the Moon's surface.

Craters, craters, everywhere

The craters on the Moon come in all different sizes, and the biggest ones are even visible to the naked eye – just! But through binoculars or a telescope, you can see hundreds – thousands! – of craters.

Set phases to 'crescent'

Turning a small telescope – or even a pair of binoculars – on a crescent Moon reveals a beautiful sight: mountain ranges standing next to fields of craters blasted on the Moon's surface by rocks from space, and long shadows making everything stand out in 3D.

Craters all have a similar round shape. When viewed near the terminator of the Moon, the shadows cast by the Sun reveal the steep crater walls and perhaps a central bulge. You may also see rays stretch away from the crater across the Moon's surface. These rays, called *ejecta,* are usually brighter than the surrounding Moon rocks.

A lunar eclipse

A *lunar eclipse* is the cousin to a solar eclipse (see the earlier section 'Catching a glimpse of a solar eclipse') and is much less dramatic, but it still makes a great sight for Moongazers. As the full Moon passes into the Earth's shadow, the Moon can appear to darken and then change colour, turning a dark red.

Lunar eclipses aren't very common, but two things make them easier for you to observe than solar eclipses:

- ✔ When a lunar eclipse happens, it's visible from anywhere on the night side of the Earth.
- ✔ Lunar eclipses are safe to observe with your eyes, binoculars and telescopes.

You can still see the Moon during the depths of a lunar eclipse, even though it's entirely in the Earth's shadow. The Sun's light gets bent (*refracted*) through the Earth's atmosphere. The red light gets bent the most, and so that's the light that illuminates the Moon's surface during a lunar eclipse. What you see is the light from every sunrise and sunset on Earth lighting up the Moon's surface!

Viewing Planets with Your Naked Eye

Have you ever seen another planet? The chances are you have, even if you didn't know what you were looking at. You know that really bright star you saw last night? It may actually have been a planet.

The planets aren't always visible. As they orbit the Sun, the Sun's glare sometimes hides them, or they may be visible only at certain hours of the night.

The farther a planet is from the Sun in its orbit, the slower it moves against the background stars, and so the longer it stays in one constellation. Super-quick Mercury is visible for only a few days each time before vanishing behind or in front of the Sun, while stately Jupiter and Saturn may be visible for entire seasons.

The planets, in order of closeness to the Sun, are:

- Mercury
- Venus
- Earth
- Mars
- Jupiter
- Saturn
- Uranus
- Neptune

Identifying and observing the planets is a really rewarding part of stargazing. If you own even a small telescope, some wonders are visible to you within your own solar system.

The following sections describe these planets in more detail. The last two planets aren't naked-eye planets; you need special equipment to view them.

The five naked-eye planets have been known about for millennia. After all, you don't need any special equipment to see them – just your eyes will do. But beyond the orbit of Saturn are other planets. These planets had to wait to be discovered after the invention of the telescope, and even then, it took many years to track them down. These outer planets – Uranus and Neptune – are very far from the Sun, and so very faint and hard to find. It's worth persevering, though; if you manage to track down Uranus and Neptune, you'll be in a very select group of people ever to have seen them.

Quicksilver Mercury

The smallest of the planets that orbit the Sun (and the closest one), Mercury is very elusive and hard to spot. It isn't always visible, and you can see it in the sky only just after sunset or just before sunrise, almost lost within twilight glow.

Because Mercury is so close to the Sun, it whizzes around it, completing one orbit every 88 days. As a result, Mercury appears to move very quickly in the sky against the fixed stars. Astronomers have long observed this fast motion of Mercury. In fact, Roman astronomers named the planet after the messenger god, Mercury.

Mercury is best seen at what's called *maximum elongation*, when it's at its farthest point from the Sun in the sky. Because Mercury whizzes around the Sun so fast, maximum elongation occurs six times each year, three after sunset and three before sunrise. You can find the timings of maximum elongations of Mercury over the next few years in Appendix A.

Because Mercury is never very far from the Sun, be very careful observing Mercury through a telescope or binoculars, and make sure that you wait until after the Sun has set or stop well before the Sun is due to rise. Because Mercury is so small, you probably won't be able to see any features on its surface except through a very large telescope, but you may make out its disk shape and its phases like the Moon's.

Venus, the beauty

Like Mercury, Venus is only ever seen at sunset or sunrise, but unlike Mercury, it's not hard to find! Venus is the brightest object in the sky after the Sun and the Moon, and if it's above the horizon, you can't mistake it.

Venus is farther from the Sun than Mercury and so orbits the Sun more slowly, once every 225 days. Because of this distance, you see Venus moving much more slowly against the fixed stars in the sky. For a few weeks, you may see Venus in the sky after sunset, and then it'll vanish into the glare of the Sun only to reappear a few months later in the predawn sky.

As the brightest and most brilliant of all the planets, Venus was named after the Roman goddess of beauty – ironic really, given that its surface is hot enough to melt lead, its atmospheric pressure is a crushing 90 times that at sea level on Earth, and its atmosphere contains clouds of corrosive sulphuric acid!

A telescope can show the disk of Venus well, and you may also be able to observe the fact that Venus has phases, just like the Moon. Depending on where it is in its orbit around the Sun, you may see a crescent Venus, a half Venus or even a gibbous Venus.

Blood-red Mars

Mars, the Red Planet, is the next stop out from the Sun after the Earth. Mars appears as a bright, reddish star in the sky.

Mars is named after the Roman god of war, due to the god's association with blood, but it's actually red because of rust in the rocks on its surface, a much more mundane explanation.

Seen through a large telescope, Mars appears to be a disk with darker and lighter patches, and perhaps even the ice caps at its poles. The ice here, though, is not just frozen water; Mars has frozen carbon dioxide gas, too, better known as *dry ice*. You'll need a large telescope to see these features, though. Through binoculars, you'll just see a small red dot, barely recognisable as a disk.

Jupiter, king planet

The largest planet in the solar system is also, for me, the most rewarding to observe through a telescope. While Jupiter doesn't have the prominent rings you see on Saturn, it's got some amazing sights to show you.

Jupiter appears as a disk in a small telescope. If you've got good observing conditions – and a lot of patience – you may be able to pick out the bands of cloud that circle this gas giant planet. They appear as parallel grey stripes on the disk, alternating lighter and darker.

You may also see the Great Red Spot, a giant storm on Jupiter that's been raging for centuries. The Earth would easily fit inside this storm. Although it's called the Great Red Spot, it appears only as a very faint grey patch through a telescope, and you may not be able to see it, depending on which way the planet is facing.

Jupiter spins about its axis once every ten hours, and so sometimes the spot is on the far side of the planet, but only a few hours later, it will reappear.

While the planet Jupiter is interesting enough to look at, the real gems are the Galilean moons, the four biggest of Jupiter's family of 66 moons:

- Io
- Europa
- Ganymede
- Callisto

These moons appear as tiny specks in a line. Sometimes all four are visible on one side of Jupiter, while sometimes you see three on one side and one on the other, and sometimes two and two are visible. Occasionally, one or more of the moons passes in front of or behind Jupiter, so you won't see four of them.

Seeing one of the moons disappear behind Jupiter or reappear is exciting to observe. If you're very lucky, you may be able to see one of them pass in front of Jupiter and cast a shadow on the planet's surface.

You don't need a telescope to see the Galilean moons; a good pair of binoculars held steady on a tripod will do the job, too.

Ringed Saturn

The jewel of the solar system, Saturn, has rings that make it one of the most recognisable – and most beautiful – of the planets. The rings, made from billions of tiny bits of ice and dust, probably from a moon that was ripped into tiny fragments by Saturn's gravity, encircle the planet, and you can see them even with a small telescope or a good pair of binoculars.

Through small telescopes, the rings look like two ears sticking out on either side of the planet. Through a larger telescope, you can see them as rings and notice some gaps in the rings, as well as the gap between the rings and the planet.

Saturn is a gas giant like Jupiter. However, unlike Jupiter, Saturn has a rather featureless surface, and you probably won't see much detail on it, even through a big telescope.

Saturn has over 60 moons, the largest of which is Titan and is visible through a good telescope as a tiny speck of light next to Saturn.

Rolling, rolling, rolling: Uranus

Although Uranus was discovered in 1781, it was originally mistaken for a comet because of its motion. Uranus orbits the Sun out beyond the orbit of Saturn, and so is very faint. Nevertheless, even in light-polluted skies, you can find Uranus through binoculars.

If you look through a large telescope, Uranus will show a pale blue featureless disk, which may look darker around the edges. (Astronomers call this effect *limb darkening*.) The disk is tiny, though – only one-tenth the size of Jupiter seen through the same telescope.

Uranus orbits the Sun with an axis tilted at 98 degrees from the vertical. This probably came about due to a collision with another large planet when Uranus was forming, billions of years ago. Whereas most planets are thought to resemble spinning tops orbiting the Sun, think of Uranus as a ball rolling around the Sun.

Out on the edge: Neptune

The most distant planet from the Sun, Neptune is also the hardest to find. As a result, astronomers didn't discover Neptune until 1846.

Through a telescope or a large pair of binoculars, Neptune looks like a tiny blue dot, half the size of Uranus and smaller than any other planet looks through the same telescope. Due to its size, seeing any features on Neptune, except through very large telescopes, is impossible.

Keeping Track of Small Wanderers

After you track down all the planets in the solar system, you may want to start on the smaller dwarf planets and other small chunks of rock and ice floating out in space. These small wanderers will push your skills as a stargazer, but nothing is as rewarding as finally snaring that elusive asteroid or comet.

Plutoids

Poor old Pluto! It was discovered in 1930, when it became the ninth planet in the solar system, only to be demoted to dwarf planet in 2006. Pluto is now considered to be part of a family of objects out beyond the orbit of Neptune, known under a variety of names: trans Neptunian objects (TNOs), Kuiper Belt objects (KBOs), or my favourite, plutoids.

The largest plutoids are:

- ✔ Pluto (discovered in 1930; the original plutoid)
- ✔ Eris (discovered in 2005; the largest plutoid)
- ✔ Haumea (discovered in 2004; only one-third the mass of Pluto)
- ✔ Makemake (discovered in 2005; smaller than Haumea, but has a brighter surface so is easier to find)

You can see the plutoids only with a large amateur telescope. Pluto's the easiest to see, but Makemake may be visible, too, if you manage to get your hands on a big enough telescope. Even through big amateur telescopes, though, plutoids show up as dots rather than disks.

Dodging asteroids

You may have heard of the *asteroid belt,* which is a ring of chunks of rock and ice that forms a belt around the Sun between the orbits of Mars and Jupiter. But forget the sci-fi staple of flying through an asteroid field, dodging them as you go: the asteroid belt is actually very sparsely populated, and finding an asteroid in it can be pretty tricky.

Only the biggest and brightest asteroids are visible to star-gazers, but you'll be able to see some of these through tele-scopes. The brightest ones are:

- Ceres (discovered in 1801; the brightest asteroid).
- Vesta (discovered in 1807; the second-brightest asteroid)
- Pallas (discovered in 1802; the third-brightest asteroid)
- Iris (discovered in 1847; the fourth-brightest asteroid)

Through a telescope, you can make out these asteroids as tiny specks against the background stars, but it's hard to know that you're looking at asteroids and not stars – they won't stand out very clearly. But go out and look for the asteroids night after night, and you'll see that they are indeed wanderers.

Every so often, a lump of rock hurtles past the Earth much more closely than normal, and you'll be able to observe a *near Earth asteroid.* These objects are so close that they move very quickly across the sky and may be visible only through bin-oculars or small telescopes, and for a few nights at most.

Comets: Dirty snowballs

One of the most amazing sights you'll ever see in the sky is a big, bright comet, with its long tale stretching for hundreds of millions of miles through space. Big comets are rare events, though. There have been two great comets this century:

Comet McNaught in 2007, and Comet Lovejoy in 2011, which were so bright they were even visible in cities.

The body of a comet, the *nucleus,* sits like a tiny dirty ball of ice at the front of the comet, surrounded by a cloud of dust and gas called the *coma.* Comets' tails – the bits that make them so dramatic – are also made of this gas and dust.

Believe it or not, comets have two tails:

- The **dust tail** is the stream of dusty particles left behind along the path of the comet.

- The **ion tail** is made up of charged gases that are pushed out into a tail by the *solar wind,* which is a stream of charged particles streaming out of the Sun.

Astronomers can predict the appearance of some comets, because they return regularly to the skies after orbiting the Sun. Halley's Comet, one of the most famous of all comets, returns every 75.3 years. It last appeared in the skies in 1986, so it'll be back in 2061. Mark your calendars!

You may be surprised by an unexpected comet: McNaught and Lovejoy were discovered only weeks before they appeared in the sky. So next time you hear of a bright comet in the sky, make sure you have a look. They truly are once-in-a-lifetime events.

The Sky Is Falling In: Meteor Showers

One of the most amazing sites for stargazers is a *meteor shower,* when perhaps hundreds of shooting stars streak across the sky in a night. These shooting stars – or to give them their proper name, *meteors* – are fleeting visitors to the skies, each one lasting only a fraction of a second. Blink and you'll miss them!

On any clear night under dark skies, you'll see a few shooting stars. These tiny particles of rock move through the atmosphere overhead, burning up as they go. Most of these particles of rock are just floating around the Sun, minding their own business, and then the Earth comes along and hoovers them up. These particles move so quickly through the atmosphere – up to 40 miles *every second!* – that they heat up and glow, forming shooting stars – not stars at all, but bits of space rock.

Observing shooting stars is simple enough on any evening: just find a dark site with clear skies, look up and wait. But to maximise your chances of seeing one, go out during a meteor shower. The best bit of kit you can bring with you when meteor watching is a reclining deck chair – stargazing in comfort!

Ones to watch

Meteor showers happen predictably at the same time every year. They happen because large clouds of rock and dust scatter around the solar system, and some of them lie in the Earth's path. As the planet orbits the Sun, it encounters these clouds at the same point in every orbit – and so at the same time every year. When the Earth passes through such clouds of dust, the rate of meteors can shoot up, and you'll maybe catch dozens every hour.

Some of the most reliably active meteor showers every year are:

- ✔ **Perseids** in mid-August
- ✔ **Geminids** in mid-December
- ✔ **Eta Aquarids** in early May
- ✔ **Quadrantids** in early January

Great balls of fire

Sometimes a larger bit of rock from space can pass through the atmosphere, and rather than seeing a fleeting streak of light in the sky, you may be treated to a *fireball*. Fireballs can last for many seconds, and they're brighter than anything else in the sky except for the Sun and the Moon. You can often see bits of the fireball breaking off and forming their own trails. Bright ones are very rare; the brightest one I ever saw was in Cherry Springs State Park in Pennsylvania, USA, and it lit the ground up like someone had turned on their headlights! Some fireballs are even bright enough that they've been seen in daylight.

Holes in the ground

If a bit of space rock makes it to the ground, it can form a crater, but these craters are very rare on Earth. Earth's atmosphere causes most space rocks to burn up before they hit the ground, and any craters that are formed are eroded away by water and weather. The Moon is covered in craters because it doesn't have a protective atmosphere.

Viewing Manmade Lights

Believe it or not, you can also see manmade lights up in space, too. Don't worry; the astronauts haven't left the lights on. What you're seeing is sunlight reflecting off the bodies or solar panels of satellites and space craft.

Currently, hundreds of working satellites are in orbit around the Earth, along with thousands of defunct satellites, bits of old rocket and space junk.

On a clear dark night, you can see a few of these satellites pass overhead. They move rather sedately against the background stars, passing across the sky in a few minutes, and most of them are really faint, at the limits of your vision. Sometimes, though, they can get really bright.

You can find out which satellites are passing over your head on websites such as www.heavens-above.com.

The International Space Station

The International Space Station (ISS) is the biggest manmade object ever to orbit the Earth, weighing in at 450 tons. The ISS is home to six astronauts at any one time. Due to its size (it wouldn't fit on top of a football pitch), the ISS has a huge area that can catch the sunlight and reflect it back down to Earth, making it one of the brightest manmade objects visible from Earth.

As the ISS passes, you see it appear as a bright star moving slowly across the sky. The ISS is visible for only a few minutes each time, and at its brightest, it can glow brighter than the planet Venus. The ISS isn't visible every night, but some nights you can see it several times, because it orbits the Earth once every 90 minutes. You can see the ISS only at twilight, before dawn or after dusk.

Iridium flares

The ISS (see preceding section) isn't the brightest manmade object up there, though. The very brightest ones visible are called *iridium* satellites, and they appear bright for such a short time that when astronomers see them from Earth, they call them *iridium flares*.

Iridium flares can be much brighter than Venus, the brightest planet. They're so bright that you can actually see them in the daytime, if you know where to look.

Chapter 8

The Constellations

- -

- -

*O*n any clear night, you can look up and see the sky studded with stars. Your first night stargazing may seem overwhelming, because the patterns of stars that you see joined up in star maps aren't always that easily matched to stars in the night sky, but don't give up. Practice makes perfect, and after you can regularly identify one or two constellations, you can quickly get the hang of finding the others.

Stargazers use a total of 88 constellations, so on any given night, you can see about half that number. If you can identify just three or four of these constellations, you're way ahead of most people on the planet. After you can identify ten or more, the sky will start to become populated with these mythical gods, monsters and heroes from ancient legend.

Joining the Dots

Human brains are very good at pattern spotting. Whether it's seeing animals in the shapes of clouds, or faces in the rocks of nearby mountains, people seem to be naturally inclined to see familiar shapes where none exist.

Stargazers the world over have for millennia looked up at night and joined the dots to make recognisable patterns called *constellations. Constellation spotting* is the process of finding patterns in the seemingly random dots in the night sky. Naturally, the brighter stars form patterns more easily than the dimmer ones, and that's exactly what stargazers have done for thousands of years.

When you're outside trying to join the dots under a very dark sky, the sheer number of stars overhead makes the task difficult, with the thousands of faint stars crowding out the brighter ones. To make spotting the constellations easier, squint your eyes almost closed until you can only see a few stars; these stars are the brightest and make up the familiar constellation patterns. After you find these stars, open your eyes fully to see how many other faint stars appear in each constellation.

Stargazers living under light-polluted skies are able to see only the brightest few hundred stars, which sometimes makes finding the constellations – the brightest ones, at least – much easier, because you don't have thousands of extra faint stars crowding around and confusing you.

The ancient Greek skies

The ancient Greeks were keen stargazers and peopled the night skies with characters from their mythology, from Orion the hunter to Ursa Major and Minor, from Pegasus the winged horse to Hercules the warrior. Many Greek myths are played out among the stars, and the nightly light show overhead must have been a great way to teach people about these legends.

Because Rome and later western Europe adopted and adapted some Greek culture, astronomers have inherited the Greek constellation names, and most constellations used today have their origin in Greek myth.

The Greek astronomer Claudius Ptolemy did the most to catalogue the constellations. In his book *Almagest,* he lists a total of 48 constellations.

Clash of the Titans

One famous Greek myth involves a vain queen called Cassiopeia, wife of King Cepheus and mother of the princess Andromeda. Cassiopeia angered the sea god Poseidon, who sent a sea monster, Cetus, to ravage her kingdom. To appease the gods, Cepheus and Cassiopeia decided to sacrifice their daughter Andromeda to Cetus, and chained her up by the shore, hoping that by making this sacrifice they could save their kingdom. Luckily for Andromeda, a passing hero called Perseus (fresh from slaying the Gorgon Medusa, from whose dead body sprang the winged horse Pegasus) heard of her plight and slayed the sea monster, thus rescuing the princess.

Cassiopeia, Cepheus, Andromeda, Perseus and Pegasus are all constellations, gathered together in the night sky next to each other, and so this story can be told amongst the stars on any clear night.

However, the ancient Greeks couldn't see some parts of the sky from their location in southern Europe, and so Greek constellations only partially cover the sky. The other 40 constellations were named much later by European explorers who arrived in the far south and saw these new constellations for the first time. These constellations have more modern names, such as the Clock, the Sextant or the Telescope.

Constellations around the world

Of course, every culture in the world has its own myths and legends and had different ways of joining up the stars into patterns. The ancient Greek constellation names were passed on only by historical accident, and it's these names that are the formally recognised astronomical constellations.

Chinese astronomy has 31 constellations (called *enclosures*), while the Native Americans also drew patterns in the stars, as did Australian Aborigines, the ancient Egyptians and cultures the world over.

In 1922, the International Astronomical Union (IAU) finally fixed the 88 constellations and their boundaries, and stargazers around the world use these constellations. Having a defined set of constellations helps prevent possible confusion when astronomers in one part of the world refer to a constellation as looking like a queen while others see that group of stars looking like a camel!

You can find descriptions of all 88 constellations in Part III, along with details of the stars and faint fuzzies that make up these patterns.

When is a constellation not a constellation?

Quite a few famous patterns in the sky aren't one of the 88 constellations, including the:

- ✔ Big Dipper, or Plough
- ✔ Little Dipper
- ✔ Summer Triangle
- ✔ False Cross
- ✔ Teapot

Stargazers call these nonconstellation patterns *asterisms,* and they shouldn't be confused with the official constellation patterns.

The Big Dipper, shown in Figure 8-1, is one of the most famous asterisms in the sky – so famous, in fact, that most people think it's a constellation in its own right. Actually, the Big Dipper is only part of a bigger constellation, Ursa Major, the Great Bear.

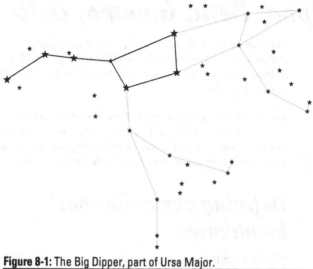

Figure 8-1: The Big Dipper, part of Ursa Major.

Table 8-1 lists some of the more famous asterisms and the constellations that you can find them in.

Table 8-1 Famous Asterisms and Their Constellations

Asterism Name	Constellation
Big Dipper	Ursa Major
Little Dipper	Ursa Minor
W	Cassiopeia
Sickle	Leo
Summer Triangle	Made up of the three brightest stars in the constellations Cygnus, Lyra and Aquila
False Cross	Made up of two stars in Vela and two in Carina
Teapot	Sagittarius

Alpha, Beta, Gamma, Delta

Stargazers identify stars and other objects in relation to which constellation they're in. No matter what you're looking for in the night sky, you can find it in one of the 88 official constellations.

Star maps or astronomy guidebooks often list the names of stars in a seemingly baffling array of numbers, letters and abbreviations. If you're going to learn your way around the sky, it helps to know how these names are given.

Defining constellations' boundaries

The 88 constellations are all given three-letter abbreviations that help identify them. Stars that lie within a constellation boundary are referred to with these three-letter codes. In addition, stargazers use what's called the *genitive* form of the constellation name. For example, the genitive form of Orion is *Orionis,* so if you're referring to a star in Orion, you would call it 'something Orionis'.

The brightest stars in any constellation are usually the ones that are used to make up the pattern, and many of the bright stars stand out so much that they've been given proper names. Regardless of whether a star has a proper name or not, though, the bright ones all have Greek letters associated with them.

Bayer letters

The brightest star in any constellation is usually given the designation *alpha,* represented by the Greek letter α, while the second brightest is *beta,* β, and so on. These Greek letters are known as the *Bayer letters.*

You can usually tell how bright a star is relative to the others in the same constellation by comparing Bayer letters: those stars with a Bayer designation near the beginning of the Greek alphabet are brighter than those who languish at the tail end of the alphabet.When a star is bright enough to be given its own Bayer letter, that Greek letter is added as a prefix to the three-letter constellation abbreviations. So if you see a star named as α, many times it's the brightest one in that constellation. (Actually, this generalisation isn't always the case. Orion makes a good illustration of an exception to the rule: in Orion, the brightest star, *Rigel,* or β Orionis, is a little brighter than *Betelgeuse,* α Orionis!)

It's not really possible to compare the brightness of stars between constellations using their Bayer letters, because the system relates to the relative brightness of stars within a constellation. For example, the brightest stars in Orion are α Orionis and β Orionis, and the dimmer ones include stars with middling Bayer letters, such as μ Orionis. But μ Orionis is brighter than the nearby α Monocerotis, the brightest star in the constellation Monoceros!

Proper names

Some of the brightest stars have a proper name, which you can use interchangeably with the star's Bayer designation. For example, the brightest star in the constellation Carina is called *Canopus,* and it also has the Bayer designation α Carinae.

The names given to stars come from a variety of different cultures and languages, but most stars have proper names from Latin, Greek or Arabic. Many of the stars in the southern hemisphere lack proper names because they weren't visible to the Greek and Arabic stargazers who catalogued the stars centuries ago. That's not to say that people in the southern hemisphere didn't name their own stars, but rather that astronomers use a rather restricted list of named stars, those formally recognised by the IAU.

The Greek alphabet

Here are the letters of the Greek alphabet, in order:

			Delta, δ	Mu, μ	Upsilon, υ
			Epsilon, ε	Nu, ν	Phi, φ
Alpha, α	Iota, ι	Rho, ρ	Zeta, ζ	Xi, ξ	Chi, χ
Beta, β	Kappa, κ	Sigma, σ	Eta, η	Omicron, o	Psi, ψ
Gamma, γ	Lambda, λ	Tau, τ	Theta, θ	Pi, π	Omega, ω

The ten brightest stars in the sky all have proper names, and they are, in order of brightness:

- ✔ *Sol*, our Sun
- ✔ *Sirius* (α Canis Majoris)
- ✔ *Canopus* (α Carinae)
- ✔ *Rigil Kentaurus* (α Centauri)
- ✔ *Arcturus* (α Boötis)
- ✔ *Vega* (α Lyrae)
- ✔ *Capella* (α Aurigae)
- ✔ *Rigel* (β Orionis)
- ✔ *Procyon* (α Canis Minoris)
- ✔ *Achernar* (α Eridani)

The not-so-bright stars

Constellations have a lot more stars than the Greek alphabet has letters, so once all the brighter ones have been given a Bayer letter, astronomers have to use a different system for naming stars. As a result, all stars within a constellation are given a number too.

Stars within a constellation are numbered from west to east (right to left in the northern hemisphere, and left to right in the southern). This number is called the Flamsteed number. The bright stars get numbered, too, so they have both a Bayer designation and a Flamsteed number.

For example, if you're looking for the star called *Deneb* in the constellation of Cygnus the Swan, it may be listed as α Cygni (its Bayer designation) or 50 Cyg (its Flamsteed number).

Very faint stars – those that require binoculars or a telescope to see – are listed with a catalogue number. Many different star catalogues and different designations exist, but two of the more common ones are the Hipparchus (HIP) catalogue and the Henry Draper (HD) catalogue.

Variable stars

Some stars vary in brightness, brightening and then dimming again. To help make these stars stand out in a star chart, some of them are given a capital letter followed by their constellation's genitive. For example, T Tauri is a variable star in Taurus the Bull.

The letters start with R and proceed to Z before starting again at RR, RS, RT and so on up to ZZ, then AA to AZ, and finally BA to BZ.

Variable stars that are bright enough to have a Bayer designation aren't given a new letter, so this system applies only to fainter stars.

Faint fuzzies

As well as stars, constellations are home to the faint fuzzies, fixed against the background of stars. Several catalogues list the faint fuzzies:

- ✔ Messier Catalogue, using M numbers, such as M45, the Pleiades
- ✔ New General Catalogue, using NGC numbers, such as NGC 4755, the Jewel Box cluster
- ✔ Index Catalogue, using IC numbers, such as IC 2602, the Southern Pleiades
- ✔ Caldwell Catalogue, using Caldwell C numbers, such as C6, the Cat's Eye Nebula

Star Hopping

Some (actually most!) of the 88 constellations are rather difficult to find and are indistinct. To help you find them, you can use the brighter constellations as signposts.

Coming up with your own signposts is a great way to navigate the sky. Don't worry if your signposts are different from the more famous ones; if it works for you, it's not wrong.

The Big Dipper as a signpost

One of the great signposts in the northern hemisphere sky is the Big Dipper, also known as the Plough, an asterism in the constellation of Ursa Major. If you can find the Big Dipper, then you're well on your way to finding lots of other constellations. 'Dipper' is an American word for a large ladle for scooping liquid. The Big Dipper asterism has three stars in a curve representing the handle, and four stars at the end of the handle representing the scoop.

Figure 8-2 shows the Big Dipper and the various constellations that you can find using it as a signpost. Here's what you can find using the Big Dipper:

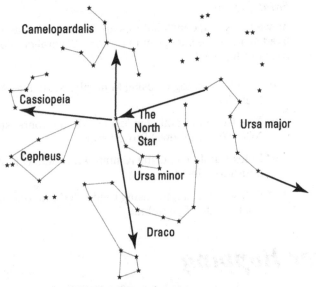

Figure 8-2: The Big Dipper as a signpost.

✔ **Pointers to Ursa Minor:** The two brightest stars in Ursa Major, α Ursa Majoris and β Ursa Majoris, lie at one end of the Big Dipper asterism. If you picture this dipper lying down, with the base of the scoop on the 'floor', draw a line up from β Ursa Majoris through α Ursa Majoris and keep going until you get to the next bright star, which is *Polaris*, α Ursa Minoris, the North Star. *Polaris* lies at the end of the tail of Ursa Minor and the end of the handle of the Little Dipper asterism. The rest of Ursa Minor curves from the North Star towards the four stars in the scoop of the Big Dipper.

✔ **Carry on to Cassiopeia:** Draw a line from α and β Ursa Majoris through α Ursa Minoris, the North Star, and go the same distance again on the opposite side of Ursa Minor, and you arrive near the distinct zigzag shape of Cassiopeia (which often looks like a W, an M or an E shape, depending on which position the constellation is in when you're observing it).

✔ **Little Dipper pointers to Draco:** After you find the North Star using the pointers of the Big Dipper, and once you've found the asterism of Ursa Minor, then lying between the Big and Little Dippers is the curve of stars making up part of the constellation Draco. Locate the head of Draco using the four stars that make up the scoop of the Little Dipper. Imagine the Little Dipper lying down, with the base of the scoop on the 'floor', with its handle rising up and to the left. The head of Draco sits 'below' the scoop.

✔ **Arc to *Arcturus*:** Go to Ursa Major, to the handle of the Big Dipper. The three stars in this handle form an arc, and if you continue this arc, it will lead you to the bright orange star *Arcturus*, α Boötis. This pointer is easy to remember: just arc to *Arcturus*.

✔ **Spike to Spica:** After you arc to *Arcturus*, continue in that direction in a straight line, driving a spike down to the next bright star, *Spica*, α Virginis. So remember arc to *Arcturus* and then drive a spike to *Spica*.

✔ **Little Dipper arcs to Camelopardalis:** Another signpost in this part of the sky uses the tail of Ursa Minor, the handle of the Little Dipper, to arc to Camelopardalis. It's not nearly as catchy as arcing to *Arcturus*, but Camelopardalis is a very faint constellation that's hard to find, so having a signpost is very handy.

Orion as a signpost

Orion is another great signpost constellation, perhaps the very best, because you can use the stars of Orion to find seven other constellations immediately around it (see Figure 8-3).

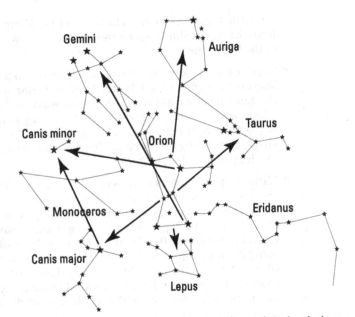

Figure 8-3: Orion as a signpost, as seen from the northern hemisphere.

The objects you can find using Orion as a signpost are:

✔ **Orion's Belt to Canis Major:** Picture Orion as a hunter standing upright, with the stars *Betelgeuse* and *Bellatrix* marking out his shoulders, and *Saiph* and *Rigel* marking out his feet. You can then follow the three stars of Orion's Belt down and to the left to find the bright star *Sirius* (α Canis Majoris) in the constellation Canis Major, the Big Dog.

✔ **Orion's shoulders to Canis Minor:** Draw a line from *Bellatrix* through *Betelgeuse* and keep going to find a solitary bright star, *Procyon*, α Canis Minoris in the constellation Canis Minor, the Small Dog. *Procyon* is one of the brightest stars in the sky, so it stands out in this blank patch to Orion's left.

✔ **Rigel and Betelgeuse to Gemini:** Draw a line from *Rigel* up past *Betelgeuse* and keep going until you reach two bright stars shining side by side. These stars are *Castor* and *Pollux*, α and β Geminorum, in the constellation Gemini the twins.

- **Above Orion's head to Auriga:** Travelling up from Orion's Belt between his two shoulders and past his head, you'll soon come to the bright star *Capella,* α Aurigae, in the constellation Auriga the charioteer.

- **Orion's Belt to Taurus:** Going back to Orion's Belt, your next target lies up and to the right, following the line of the belt to the bright star *Aldebaran,* α Tauri, in the constellation Taurus the Bull. If you keep on following this line, a short distance later you'll reach the Seven Sisters, or the Pleiades, star cluster.

- **Orion's right leg to Eridanus:** Just to the right of Orion's right foot, *Rigel,* you'll find the first few dim stars in the huge meandering constellation of Eridanus the River. The rest of this constellation snakes down a great distance, ending at the bright star *Achernar,* not visible from the northern hemisphere.

- **Beneath Orion's feet to Lepus:** Directly beneath Orion's feet, you'll find the faint constellation of Lepus the Hare.

- **Between Canis Major and Minor to Monoceros:** Another faint constellation lies in the seemingly blank part of the sky between Canis Major and Canis Minor. If you draw a line from *Sirius* to *Procyon,* then you'll be passing through Monoceros.

The Southern Cross as a signpost

You can find a lot of great signposts in the southern skies, too, none more so than the Southern Cross, Crux. The constellations that you can find by using the Southern Cross as a signpost are shown in Figure 8-4.

Picture Crux as a traditional Christian cross, with the long axis running up and down and the short axis running across this, higher up than the centre line of the long axis.

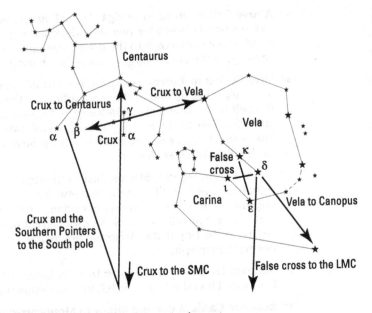

Figure 8-4: The Southern Cross as a signpost.

The objects you can find using The Southern Cross as a signpost are:

- **Crux to the southern pointers:** Draw a line from the star δ Cruxis, through the star β Cruxis. This line points to two very bright stars side by side, α and β Centauri, which are known as the southern pointers.

- **Crux and the southern pointers to the South Pole:** If you want to find north in the northern hemisphere, then you can use the pointer stars of the Big Dipper to find the North Star, *Polaris*. In the southern hemisphere, things get a bit trickier because no south star marks south for you. Instead, you should use two lines: the first line is drawn from γ Cruxis through α Cruxis and extended in that direction. The second line is drawn bisecting α and β Centauri and continuing perpendicularly away from these two stars. Where the first line meets the second line is due south.

- ✓ **Crux to Centaurus:** You've already found the two brightest stars of the constellation Centaurus the Centaur, α and β Centauri, which make up the centaur's front legs. To find the rest of this large constellation, look 'above' Crux to find the back legs, and above and to the left of the back legs to find the body and head.

- ✓ **Crux to Vela:** If you draw a line from left to right along the short axis of Crux and keep going, you eventually find the large constellation Vela the Sail.

- ✓ **Vela to the False Cross:** Two of stars at the bottom of Vela, κ and δ Velorum, make up part of the False Cross, along with two stars from the top of Carina the Keel, ε and ι Carinae. Because the False Cross looks like a slightly bigger version of Crux, it's often mistaken for its more useful namesake by stargazers trying to find south.

- ✓ **Vela to Carina:** The constellation Carina the Keel lies beneath Vela. The two stars of Vela that are part of the False Cross, κ and δ Velorum, point to the very bright star *Canopus, α* Carinae, in Carina.

- ✓ **Crux to Tucana and the Small Magellanic Cloud:** If you follow the long axis of Crux down past the South Pole, you'll eventually come to the Small Magellanic Cloud in the constellation of Tucana.

- ✓ **The False Cross to Dorado and the Large Magellanic Cloud:** Similarly, you can use two stars in the False Cross to find the Large Magellanic Cloud. Draw a line from δ Velorum through ε Carinae and keep going to the Large Magellanic Cloud in the constellation of Dorado.

Cassiopeia as a signpost

Cassiopeia is another prominent and easily identifiable northern hemisphere constellation. The brightest stars in Cassiopeia make up a zigzag of stars, an asterism that looks like a letter W or M, depending on which way up you see it. If you're not sure whether you're looking at Cassiopeia, then you can use the Big Dipper as a signpost to find it.

After you find Cassiopeia, you can use it as a signpost to find all the constellations associated with this queen (see Figure 8-5).

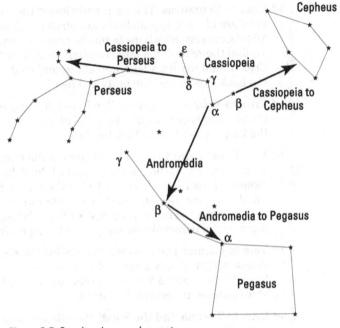

Figure 8-5: Cassiopeia as a signpost.

The objects you can find using Cassiopeia as a signpost are:

✔ **Cassiopeia to Andromeda:** After you find Cassiopeia using the Big Dipper (the Plough) – see preceding section – you can look for the obvious W shape (or M shape, depending on which way up you are). The right-hand V of this W (α, β and γ Cassiopeiae) works as an arrow that you can follow to find the bright star β Andromedae in the constellation of Andromeda the princess. β Andromedae is the centre star of three in a curve, with α Andromedae lying to one side and γ Andromedae to the other.

✔ **Andromeda to Pegasus:** α Andromedae is actually counted as one of the four stars in the Great Square of Pegasus the Winged Horse, despite the fact that it's not in Pegasus at all!

✓ **Cassiopeia to Perseus:** To find the hero Perseus, who rescued the princess Andromeda, use two stars in Cassiopeia as pointers: draw a line from the middle of the five stars in the W, γ Cassiopeiae, down through the second star from the left, δ Cassiopeiae, and continue to find the sweep of stars representing Perseus's out-stretched arm, sword in hand.

✓ **Cassiopeia to Cepheus:** The right two stars of the W of Cassiopeia point towards the rather dim constellation Cepheus the King.

Chapter 9

Mapping the Skies

. . . .

In This Chapter

▶ Using a star map

▶ Identifying the different markings on a star map

▶ Choosing your first star map

. . . .

*E*ven the most dedicated stargazers have to use star maps to find the objects they're looking for. With your naked eyes, you can see more than 6,000 stars, 88 constellations, dozens of faint fuzzies, five planets, one Moon and more. Remembering where everything is located is impossible, but a good star chart can help you find your way.

In this chapter, you find out what all the various markings on a star chart are and how you can use them to find your way around. In Part III, you find star maps for each of the 88 constellations, with the stars named and many interesting faint fuzzies marked.

Reading a Star Map

One of the most important skills you'll need to master is reading a star map. A star map helps you become familiar with the constellations and the stars and faint fuzzies in them and means that you're never at a loss for something to look for in the night sky.

Mastering star maps takes time and patience, and the more often you use one, the more like second nature it will become. At first glance, a star map may look like a bewildering array of dots, shapes and lines. But if you remember a few simple things, you'll find that everything starts to make sense.

In black and white

A star map is white with black dots on it, unlike the sky, which is black with white dots on it (not really, but that's how it looks!). This colour scheme is really just to save on ink! Printing a few black marks on an otherwise blank sheet of paper is cheaper than printing on black with a few white dots. That said, some star maps do use a more realistic colour scheme.

If you're using your map outside at night, use a dim red torch to look at it so that you don't spoil your night vision. If you use a red torch, you won't be able to see colours at all, so a simple black-and-white star map is actually better for outdoor use than a coloured one.

Big dots, small dots

Stars on star maps are represented by dots or, more correctly, circles. But how do you show how bright a star is just by using a dot on a piece of paper?

Astronomers have come up with a cunning solution: the brighter a star is, the bigger the dot used to mark its position. As a result, star maps contain a range of dot sizes in order to show bright, medium and dim stars.

The maps in Part III use six different sizes of dot to show the brightness of stars. The largest dots are the very brightest stars, which are visible no matter how bad the light pollution is. The smallest dots are the faintest stars that you'll be able to see from a rural sky without straining your eyes too much.

Some star maps use a much broader range of dot sizes, to show very faint stars that are only visible in perfect conditions or with binoculars or a telescope. These maps are great if you want to explore the sky in lots of detail, but if you're just learning your way around the constellations, less is often more when it comes to star maps.

Very simple star maps – ones that only have one size of dot and show only the brightest stars in the major constellations – can be useful, too. While this kind of map misses out a lot of the detail, it's a good way to discover signposts in the sky.

Magnitudes – a measure of brightness

Astronomers measure a star's brightness using something called the magnitude scale, where each star – or planet or faint fuzzy – has a magnitude brightness value. When looking up at stars from the Earth – which you'll do – astronomers measure how bright the star appears *to them*. This is the star's *apparent magnitude*. Because you can't tell how far away a star is just by looking at it, you don't know how intrinsically bright it is – what its *absolute magnitude* is. In Chapters 10–15 of this book you'll find the apparent magnitudes of the brightest stars in each constellation.

The brighter an object is, the lower the magnitude number. For example, the North Star has an apparent magnitude of 2.0, whereas *Betelgeuse* in the constellation of Orion has an apparent magnitude of 0.4, which makes *Betelgeuse* brighter than the North Star.

Your eye can only see stars brighter (with a lower magnitude) than magnitude 6.5 (if you've got great eyesight, and in perfect sky conditions). Very bright objects (like the bright stars *Sirius*, *Canopus* and *Arcturus*, the planets Venus, and Jupiter, or the Moon) have negative magnitudes.

The magnitude scale is not a linear scale. For every point up the magnitude scale, an object gets around two-and-a-half times dimmer (2.52 to be more exact). So a star of magnitude 1 is 100 times brighter than a star of magnitude 6 (that is, there's a difference of five magnitude points, and so a difference of 2.52 x 2.52 x 2.52 x 2.52 x 2.52 = 100 in brightness).

Naming stars

Star maps all differ in how they name stars. (See Chapter 8 for more on how astronomers name stars.) Some star maps don't name any stars. Others give only the proper names of the very brightest stars, such as *Betelgeuse, Canopus* or *Sirius*. Still others list stars by their Bayer designation and/or Flamsteed number (like α, β or γ, or 29, 30 or 31). Figure 9-1 shows how the naming of stars can vary, using the constellation of Orion as an example.

However the labelling is done, most maps name only the brighter stars to avoid cluttering up the map with too much information.

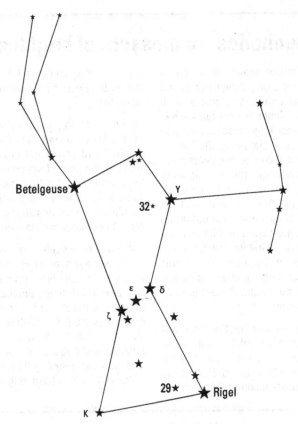

Figure 9-1: Some stars in Orion labelled with proper names, Bayer (Greek) letters, and Flamsteed numbers

Constellation lines

You can join up the brightest stars in a constellation on a star map with lines showing the general outline, or maybe just the brightest asterism within and among the constellations, such as the Big Dipper or the Summer Triangle. Although the boundaries of all the constellations are very well defined, the way you join the stars within a constellation to make a pattern may vary slightly from map to map.

Star maps may also name the constellations, too, usually in a different style to star names. For example, the star names may appear in lower case, while the constellation names are in upper case. Usually, this difference is enough, but some maps also include lines that mark out the constellation boundaries, so that you can tell which constellation every star belongs to.

Doubles and variables

Some stars are more interesting than others. For example, some stars are actually multiple stars, for which, through binoculars or a telescope, you can see more than one star making up the single dot visible to your naked eye. Some stars are variable stars that change in brightness over time, dimming and brightening. These stars are especially interesting to stargazers, so they usually get marked slightly differently than the common-or-garden stars.

You'll often find the following types of stars marked differently on a star map:

✔ **Multiple stars:** Some multiples, such as *Mizar* and *Alcor* in the handle of the Big Dipper, are two stars that are far enough apart that you can mark them both on the star map without having to use the special symbol. In some cases, these stars may overlap, and one black dot is placed on top of, or overlapping with, another black dot (see Figure 9-2). In this case, the smaller dot is usually given a white border so that you can see it better.

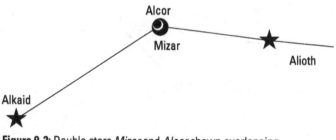

Figure 9-2: Double stars *Mizar* and *Alcor* shown overlapping.

In most cases, though, multiple stars are closer together than you can easily mark on a map, and so the star is given a slightly different label – a dot with a line through it. In the case of *Mizar* and *Alcor*, both of these stars are themselves double stars, and so they both have lines through them (see Figure 9-3).

Figure 9-3: Double stars *Mizar* and *Alcor* marked with horizontal lines showing that they are both double stars.

✔ **Variable stars:** Stars that change in brightness are sometimes marked with a dot inside a circle (see Figure 9-4). The size of the circle represents the brightest the star gets to, while the size of the dot shows the faintest it gets to. Imagine these marks with the inner black dot growing and shrinking in size, sometimes filling the whole circle, sometimes fading to the size of the small inner dot.

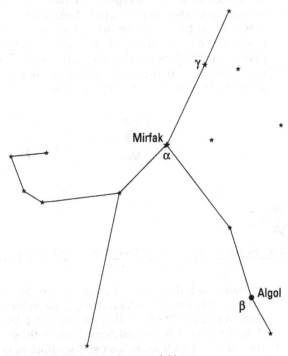

Figure 9-4: *Algol* marked as a variable star.

Faint fuzzies on the map

Different star maps use different ways of representing the faint fuzzies. In the guides in Part III, I label the more interesting ones with a + symbol.

If you want to dig deeper, consider learning the symbols that your star map uses for the different types of faint fuzzies, galaxies, open clusters, globular clusters, diffuse nebulae and planetary nebulae, which might all have their own symbols. For example, galaxies are often marked with an oval shape; the bigger the oval, the larger the galaxy looks in the sky.

The Milky Way

When the Milky Way is marked on star maps, it's shown as a faint wash of colour – maybe a grey wash on a black-and-white map. More detailed maps may have two shades of the wash, showing the brighter and the darker parts of the Milky Way.

Coordinate lines

To help with navigation, some star maps include coordinate lines that allow you to more accurately track down a faint fuzzy or an interesting object.

Chapter 9 talks about the coordinates right ascension (RA) and declination (dec) – roughly equivalent to longitude and latitude on Earth – and how we can use them to track the stars. It's these coordinates that are sometimes marked on a star map.

RA and dec (which is also represented by the Greek letter δ) are coordinates marking a star's (or galaxy's) position with respect to two points in the sky:

- ✔ The *celestial equator* (a line in the sky directly above the Earth's equator)
- ✔ The exotically named *First Point of Aries,* which is situated in the constellation of Pisces

Declination tells you how far from the celestial equator a star is, and is measured in degrees, just like latitude on Earth tells you how far from the Earth's equator you are, in degrees. When astronomers say that a star, such as *Canopus,* has a declination of –52°41', it means that it's that many degrees south (the minus sign applies to all stars in the southern hemisphere) of the celestial equator.

RA is measured as the angular distance along the celestial equator from the First Point of Aries, a position on the celestial equator in the constellation of Pisces. Rather than measure this angle in degrees, RA is measured in hours, minutes and seconds, so that the whole circle around the Earth is 24 hours. Therefore, when someone says that *Canopus* has an RA of 06^h24^m, it means that its RA coordinate is just over one quarter of the way around the sky (6^h is one-quarter of 24^h) from the First Point of Aries. RA is always measured eastward from the First Point of Aries.

Because 24^h of RA makes one complete circle (360 degrees), you can work out that 1^h of RA is equal to 15 degrees, and 40^m of RA is equal to 10 degrees. Using the simple rule that one clenched fist at arm's length is around 10 degrees, you can move through the sky in 40^m intervals.

An object may occasionally be listed in a book, magazine or online without a star map, but has its RA and dec listed. A star map with the RA and dec coordinate grid marked on it should allow you to find the object with a bit of practice.

If you have a telescope with a computerised go-to feature, then you can input the RA and dec of any object, and the telescope will automatically find it.

Other items

Your star map may also have some other items marked on it as well:

✔ **The equator:** The celestial equator is a line in the sky that lies directly above the Earth's equator. Try to imagine this line as the projection of the Earth's equator out into space. Effectively, the equator is just the zero line of declination, but it sometimes gets marked more prominently.

✔ **The ecliptic:** The ecliptic line, or the zodiac, is the line in the sky along which all the wanderers – the Sun, Moon and the planets – move. The ecliptic is the projection out into space of the plane of the solar system. Because the ecliptic is where you have to look to find planets (and other things, such as asteroids), it's often marked with its own line on a star map. You may find it marked with coordinate markings, too – say, every 10 degrees may be marked. These markings help you find an object that's listed with its *ecliptic coordinates* (instead of RA and dec).

As well as all of these lines marking the skies in your star map, you may encounter additional ones, such as *galactic coordinates*. The galactic coordinate system uses the plane of the galaxy as its baseline, with the zero point being the centre of the galaxy.

Buying Your First Star Chart

The constellation guides in Part III give you plenty of things to look for, and this book's portable nature means that you can take it out with you at night and use it to hunt down those interesting faint fuzzies or elusive constellations.

However, I include only a selection of the faint fuzzies and only the bright-ish stars, the ones you can see with your naked eye from suburban skies. Although my constellation guides give you a great start in stargazing, hopefully after you exhaust this book, you'll purchase a more detailed guide.

In the following sections, you meet the different kinds of star maps and work out which one is best for you.

Planispheres

A very useful type of star map is the *planisphere,* a round map in two parts, one of which spins on top of the other. The lower part has all the stars that are visible from your latitude, while the upper part is mainly opaque with only a small transparent window. The upper part spins around, and so the window moves showing different stars beneath it. Around the outside rim of the upper part are the hours of the day, and around the outside rim of the lower part are the days of the year.

Simply spin the upper part around so that the time when you're observing is pointing at the day when you're observing. After you rotate the planisphere, the stars in the window are the stars that you can see overhead at that instant.

You can also find the cardinal points – north, south, east and west – marked around the outside, but even a quick glance shows that east and west are the wrong way around! That's because planispheres are designed to be used held upside down above your head, with you looking up into them, not down onto them. Don't be put off if it takes you a little time to master a planisphere.

You can see different stars, depending on where you're stargazing from on Earth. This means that you'll need to make sure that the planisphere you buy is for the correct latitude. Differences of around 10 degrees are probably fine; any more than that amount and your planisphere won't show you the right stars.

A planisphere is a useful addition to your stargazing kit. Here are some advantages of planispheres over other star maps:

- ✔ They give you a view of what's overhead at that particular point in time.

- ✔ They usually contain only the bright stars and constellations, which makes finding the basic patterns easier.

- ✔ They are flat, light and portable.

However, planispheres have a few drawbacks, too:

- ✔ They're latitude specific, so you'll need to buy a new one if you want to stargaze while on holiday in a different part of the world.

- ✔ They aren't very detailed, so they won't help you track down many faint fuzzies or dimmer stars.

- ✔ They can be tricky to master the first few times you use them, especially because you have to hold them above your head to read them properly.

Monthly star maps

Many great astronomy magazines are available. Most of them are monthly, and the centre pages of many of them contain detailed star maps for that month, showing what's up overhead (see Figure 9-5). You can also find online resources that allow you to print a star map for the month.

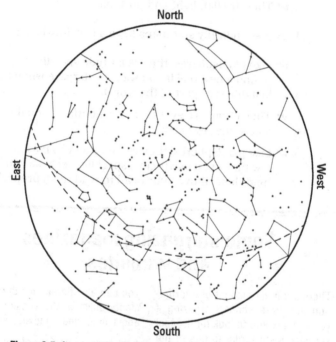

Figure 9-5: An example of a monthly star map.

Star maps operate much like a planisphere, but with no moving parts; think of a star map as a snapshot of what a planisphere shows you.

Just like with a planisphere, you have to hold star maps over your head in order to align them correctly with the sky, but they'll have more timely details, such as the position of the planets that month.

Monthly star maps are useful in a variety of ways:

- ✔ They give you a more detailed view than a planisphere can of what's overhead that month.

- ✔ Given that you can find them in topical astronomy magazines, they can include very up-to-date features, like new comets.

- ✔ They are flat, light and portable.

However, monthly star maps have a few drawbacks, too:

- ✔ Like planispheres, they're latitude specific, so you'll need to find a new one if you want to stargaze while on holiday in a different part of the world.

- ✔ They're only relevant for one month, so you'll need to get a new one regularly.

- ✔ Like planispheres, they can be tricky to master the first few times you use them, especially because you have to hold them above your head to read them properly.

Some stargazing magazines and websites

There's a huge amount of information out there about stargazing if you know where to look for it. Most stargazers subscribe to one of the monthly astronomy magazines, which have great features, articles, equipment reviews and up-to-date monthly star maps. Here are some that you might think about purchasing:

Astronomy Now (UK)

Astronomy (US)

BBC Sky at Night (UK)

Sky & Telescope (US)

You can also find a wealth of information online on the websites for the above magazines, and also at:

Astronomical Society of the Pacific www.astrosociety.org

British Astronomical Association www.britastro.org

Portal to the Universe www.portal totheuniverse.org

Society for Popular Astronomy www.popastro.com

Universe Today www.universe today.com

Guide books

Plenty of great astronomy guide books, just like this one, have constellation maps, monthly what's-up guides and background astronomy info. Guide books are a good way of learning the sky and are really useful as resource books to dip into to find new faint fuzzies or constellations.

Stargazing guidebooks can be very useful:

- ✔ They can contain much more information than a simple star map or planisphere ever could, allowing you to learn much more about what you're observing than just where and when it's in the sky.

- ✔ They have information about objects that aren't visible that month, and so you can plan your stargazing schedule well in advance.

However, guidebooks have a few drawbacks, too:

- ✔ They are often a little more expensive that a monthly magazine.

- ✔ They are not quite as portable, although smaller books like this one can easily be carried around for reference.

Star atlases

The most detailed star maps are gathered together in *star atlases,* large books full of information for the keen stargazer. At first glance, the maps in star atlases may look like a huge array of dots and marks on paper. With practice, you can begin to discover how to use star atlases effectively, and they'll become useful tools in your stargazing kit.

Star atlases are incredibly useful:

- ✔ They give you a very detailed view of all the objects visible from anywhere on Earth.

- ✔ They often include far fainter objects than those you find on a monthly star map or planisphere.

However, they do have a few drawbacks, too:

- ✔ They are often quite expensive.

- ✔ They are much larger and bulkier than more-portable guidebooks.

- ✔ They are really only useful as desktop reference books, and so you probably won't take one out stargazing with you.

Computer software and smartphone apps

There are some excellent stargazing computer programs and apps available for your desktop or laptop computers, or for your smartphone or tablet.

These programs allow you to choose your observing location and what time you want to observe, and often have huge databases of information about the stars, planets and faint fuzzies, which is available at the click of a finger.

They are useful in many ways:

- ✔ You can set them for any location on Earth (and sometimes off it!), so you can see what the sky looks like from different places.

- ✔ You can set them for any time – present, past and future – so you can see what the sky looks like years in advance or on the day you were born.

- ✔ They have huge amounts of searchable objects, so you can almost always find what you're looking for.

However, they do have a few drawbacks, too:

- ✔ Computer-based programs can't be taken outside stargazing with you, so they're really only for planning your next stargazing trip.

- ✔ More-portable tablet and smartphone apps can be used outside, but they can harm your dark adaptation. Many of them have a 'night vision' mode that dims the display and turns it red to protect your night vision, but it will still have some negative impact.

Examples of great stargazing software

Here are some examples of great pieces of stargazing software:

Stellarium (computer software, free)

Celestia (computer software, free)

Star Walk (smartphone/tablet app)

GoSkyWatch (smartphone/tablet app)

SkyMap (smartphone/tablet app)

Part III
Star Hopping

The 5th Wave   By Rich Tennant

"That reminds me—we're having Pisces in white wine sauce tonight for dinner."

In this part...

In this part I list all of the 88 constellations visible in the sky, from the big bright obvious ones like Orion the Hunter, to the tiny faint hard-to-spot ones like Mensa the Table!

Each chapter deals with a different set of constellations. Four of the chapters (11-14) list the constellations that are best seen in specific seasons, while Chapters 10 and 15 list the stars visible round the north and south polar points respectively.

Each chapter has a star map showing all the constellations in that chapter, to help you find them more easily, as well as a detailed star map of each constellation, showing the names and designations of the brighter stars, and the interesting faint fuzzies that you can find in that constellation. You'll also find details about many of these objects, and a table of info about the brightest stars in the constellation.

Chapter 10

Northern Polar Constellations

*T*he stars above the North Pole will soon become very familiar to stargazers in northern latitudes. They include constellations like Ursa Major, where you can find the famous Big Dipper, Cassiopeia with her distinct W shape, and Ursa Minor, which contains the most famous star in the night sky, the North Star.

This chapter includes star maps and lists of interesting features for the northern polar constellations.

Northern Polar Constellation Map

The constellations in this chapter, shown in Figure 10-1, are near the North Star, so you'll only be able to see all of them if you're in the northern hemisphere. If you were at the North Pole, they'd be spinning directly overhead, but for northern hemisphere observers they spin in an anticlockwise direction around the North Star.

These constellations, when observed from northern latitudes, are *circumpolar*, meaning they never rise or set. That's very handy because it means they're above the horizon year-round and are visible all night, so you'll soon get used to finding them and using them as signposts to other constellations.

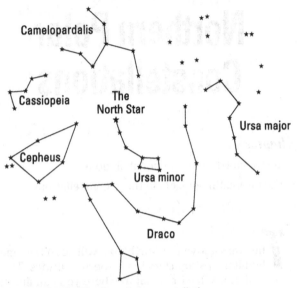

Figure 10-1: The Northern Polar constellations.

Where in the sky you have to look depends on where on Earth you're stargazing. The northern polar constellations all appear towards the north of the sky, but the closer you are to the North Pole, the higher they are. In fact, the North Star's altitude above the northern horizon is equal to the latitude you're stargazing from.

This chapter talks about these six northern polar constellations:

- ✔ Ursa Major, the Great Bear
- ✔ Cassiopeia the Queen
- ✔ Ursa Minor, the Little Bear
- ✔ Draco the Dragon
- ✔ Cepheus the King
- ✔ Camelopardalis the Giraffe

Ursa Major

Ursa Major, or the Great Bear (see Figure 10-2), is one of the best-known constellations in the whole northern sky, because it contains the pattern of seven stars called the Big Dipper, or the Plough. The stars in the Big Dipper are bright enough that you can see them from a city, and the shape they make is very distinctive – literally like a big ladle, or dipper.

The seven stars of the Big Dipper actually represent the hindquarters and tail of the Great Bear. You may be able to make out the head and legs of the bear, too.

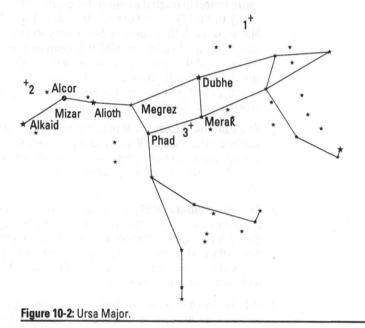

Figure 10-2: Ursa Major.

A great star to observe is the middle star of the handle of the Big Dipper. It's actually a double, or *binary,* star made up of two stars called *Mizar* and *Alcor. Mizar* is the brighter of the pair, about five times brighter than its companion. If you have good eyesight, you should be able to spot the fainter *Alcor* next to it. Finding a double star with the naked eye is unusual; to see most double stars, you need to use binoculars or a telescope.

The two stars furthest from the handle of the Big Dipper, called *Merak* and *Dubhe,* are known together as the pointer stars, because they point to the North Star, *Polaris,* in Ursa Minor. As the Earth spins, the stars in Ursa Major all spin around the North Star, but no matter what time of night you look, the pointers always point north (see Figure 2-2 in Chapter 2).

Here are some things to look for in Ursa Major. The numbers in the list correspond to the numbers in Figure10-2:

1. A pair of galaxies, **M81 and M82,** sit very close together in the sky. You can see both of them at the same time, through a medium telescope. They're a bit tricky to find (you can use the brighter stars of Ursa Major to track them down), but after you locate them, you won't mistake them. M82 is known as the Cigar Galaxy, and it really does look like a cigar-shaped smudge next to its rounder companion. M82 is what's called a starburst galaxy, because it's in the process of making lots of new stars.

2. The **Pinwheel Galaxy, M101,** is a spiral galaxy that you see face-on. You'll see it as a faint smudge through a small telescope or binoculars. Through a really big telescope, you may be able to make out its spiral arms.

3. The **Owl Nebula, M97,** is a planetary nebula (see Chapter 6) with two dark regions looking like eyes and giving it the appearance of an owl's head. Through binoculars, the Owl Nebula is very faint and small. (It looks about the size of Uranus.) Only a big telescope will show you the dark eyes.

4. **M40** is the only object in Messier's list (see Chapter 6) that isn't actually a faint fuzzy. In fact, it's just a double star, and it was a mistake to include it in the list in the first place!

5. **M108** is an edge-on galaxy in the same field of view as the Owl Nebula, M97.

6. **M109** is another spiral galaxy next to the bright star *Phad,* γ Uma.

Ursa Major info

Abbreviation: UMa

Genitive: Ursae Majoris

Best months for observing in during the evening: Visible all year from northern latitudes, but highest in the evening from February through May

Bright Stars List

Name	Bayer Designation	Type	Mag	Distance (ly)
Dubhe	α UMa	F7V	1.81	124
Merek	β UMa	A1V	2.34	79
Phecda	γ UMa	A0V	2.41	84
Megrez	δ UMa	A3V	3.32	81
Alioth	ε UMa	A0p	1.76	81
Mizar	ζ UMa	A2V	2.23	78
Alkaid	η	B3V	1.85	101
	ψ UMa	K1III	3.00	147

Cassiopeia

Cassiopeia (see Figure 10-3) is another very prominent constellation. It's not as famous as the Big Dipper (see preceding section), but it's just as easy to recognise. If you use the pointer stars in the Big Dipper to find the North Star, keep going and you get to Cassiopeia.

Cassiopeia's five brightest stars make a zigzag line in the sky. Depending on what time of night you're stargazing, Cassiopeia may look like a W or an M. The W shape of five stars is meant to represent the queen Cassiopeia sitting in her chair.

Look for these things in Cassiopeia. The numbers in the list correspond to the numbers in Figure 10-3:

Sitting in the Milky Way

The Milky Way appears to run straight through Cassiopeia, making this constellation a great one to use to find this milky grey band of light in the sky. If you're stargazing from somewhere with a lot of light pollution, the Milky Way can be hard to spot, but Cassiopeia can help you find it. In November, Cassiopeia sits high overhead late at night, with the Milky Way stretching through it in an arc from east to west.

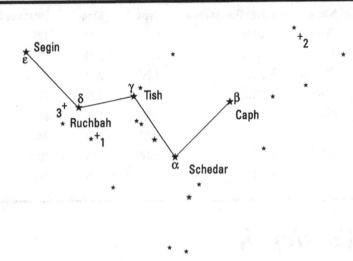

Figure 10-3: Cassiopeia.

1. The **Owl Cluster, NGC 457,** lies beneath the W of Cassiopeia. If you look at the Owl Cluster through binoculars or a small telescope, you can see that it's an open cluster of around 150 stars, with two brighter ones making the eyes of the owl.

2. Through binoculars, **M52,** another open cluster, looks like a small fuzzy patch, but through even a modest telescope you can see a rich collection of faint stars in a fan shape.

Cassiopeia info

Abbreviation: Cas

Genitive: Cassiopeiae

Best months for observing in during the evening: Visible all year from northern latitudes, but highest in the evening from October through December

Bright Stars List

Name	Bayer Designation	Type	Mag	Distance (ly)
Schedar	α Cas	K02	2.24	228
Caph	β Cas	F2III	1.17	54
Navi	γ Cas	B0IV	2.15	613
Ruchbah	δ Cas	A5V	2.66	99
Segin	ε Cas	B2p	3.35	442

3. **M103,** another open cluster, is more spread out than M52. You can easily find M103 in binoculars due to its closeness to *Ruchbah,* the second star in the W of Cassiopeia, but it'll appear only as a fuzzy patch. You need a telescope to make out individual stars, but even then it's quite hard to find, because it's so spread out.

Ursa Minor

The Little Bear, Ursa Minor (see Figure 10-4), is more famous than it really should be, thanks to its brightness and how easy it is to find, and because of the presence of *Polaris,* the North Star.

Ursa Minor's shape is quite similar to the Big Dipper's; it has three stars in a handle joining on to four stars in a rough box shape. For that reason, Ursa Minor is also known as the Little Dipper.

The brightest star in Ursa Minor sits at the end of the Little Dipper's handle: *Polaris*, the North Star. This bright star lies almost directly above the Earth's North Pole, and so as the

Earth spins about its axis once a day, *Polaris* stays fixed in the same place in the sky. That means that if you can find *Polaris* using the pointer stars of Ursa Major, and stand facing it, you'll be facing north. This method is a great way to navigate at night, and *Polaris* allowed explorers to find their way around the globe long before the magnetic compass was invented.

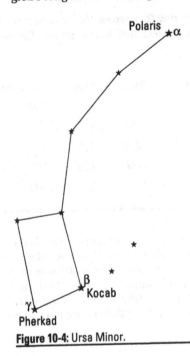

Figure 10-4: Ursa Minor.

Try it yourself! Take a map out at night and find your way around using the stars to guide you. It's a very rewarding experience.

Not the North Star forever

Although *Polaris* is currently above the North Pole, it hasn't always been there, nor will it always be, thanks to the wobbling of the Earth's axis as it spins, known as *precession*. This wobble isn't noticeable except over thousands of years. In 2500 BCE, *Polaris* wasn't the North Star; instead, *Thuban* in the constellation Draco was. Around 4000 CE, the North Star will be the star *Errai* in Cepheus.

Ursa Minor info

Abbreviation: UMi

Genitive: Ursae Minoris

Best months for observing in during the evening: Visible all year from northern latitudes, but highest in the evening from May through June

Bright Stars List

Name	Bayer Designation	Type	Mag	Distance (ly)
Polaris	α UMi	F7Ib	1.97	431
Kocab	β UMi	K4III	2.07	126
Phercad	γ UMi	A3II	3.00	480

Although you can't easily observe faint fuzzies in Ursa Minor, you can see a beautiful group of stars next to *Polaris* called the *Engagement Ring.* When you look through a telescope or a good pair of binoculars, the stars next to *Polaris* form a loop, looking like an engagement ring, with *Polaris* as the bright diamond.

Draco

Draco the Dragon (see Figure 10-5) snakes through the sky, wrapping around Ursa Minor (see preceding section) before twisting its head away again. The stars in Draco are mainly pretty faint. The brightest stars are *Eltanin* and *Thuban,* the latter sitting between the body of the Little Dipper and the Handle of the Big Dipper.

You may not see *Thuban* from a light-polluted site, as it's not especially bright. If in doubt, you can use the two stars in the body of Ursa Major that aren't the pointer stars to point towards *Thuban.*

Draco info

Abbreviation: Dra

Genitive: Draconis

Best months for observing in during the evening: Visible all year from northern latitudes, but highest in the evening from March through September

Bright Stars List

Name	Bayer Designation	Type	Mag	Distance (ly)
Thuban	α Dra	A0III	3.67	309
Rastaban	β Dra	G2II	2.79	361
Eltanin	γ Dra	K5III	2.24	148
Altais	δ Dra	G9III	3.07	100

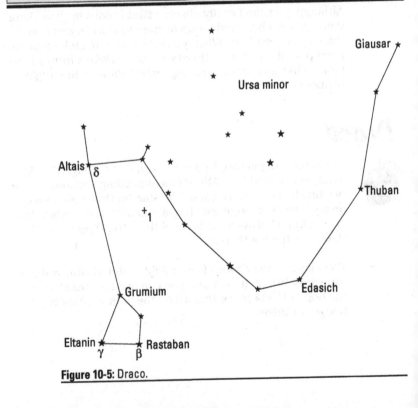

Figure 10-5: Draco.

The **Cat's Eye Nebula, NGC 6543** (numbered 1 in Figure 10-5), is the best faint fuzzy in Draco, but even still it's hard to spot. You need a big telescope to see it looking like anything more than an out-of-focus star, and you need to use powerful magnifications to see any structure in the nebula. The Cat's Eye should appear as a blue or green disk with a faint star in the middle. With a powerful eyepiece on a big telescope, you may be able to make out that the Cat's Eye has two distinct lobes.

Cepheus

Cepheus the King (see Figure 10-6) is husband to Cassiopeia and is very much the lesser constellation in the marriage. The stars of Cepheus are all quite dim, making a shape something like a square with a pointy triangular hat on.

The topmost star of this hat is called *Errai*. Between 3000 and 5000 CE, *Errai* will be the North Star, due to the precession of the Earth's axis.

Cepheus info

Abbreviation: Cep

Genitive: Cephei

Best months for observing in during the evening: Visible all year from northern latitudes, but highest in the evening from September through October

Bright Stars List

Name	Bayer Designation	Type	Mag	Distance (ly)
Alderamin	α Cep	A7IV	2.45	49
Alfirk	β Cep	B2III	3.23	595
Errai	γ Cep	K1IV	3.21	45

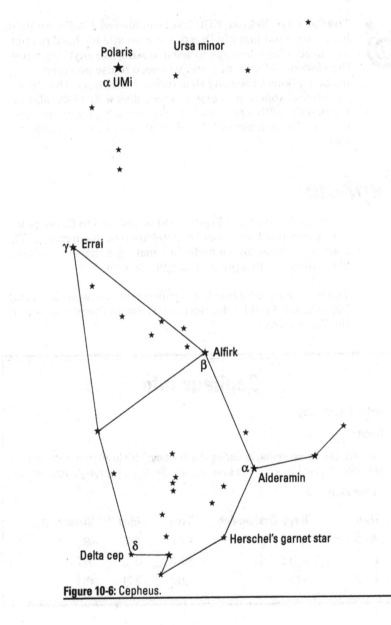

Figure 10-6: Cepheus.

Here are a couple of things to look for in Cepheus.

- ✔ **Herschel's Garnet Star,** μ **Cephei,** is the reddest star visible to the naked eye. It's quite dim, and so you can't see it from light-polluted stargazing sites. If you turn binoculars on it, then you'll see the red colour really stand out compared with the other fainter stars around it.

- ✔ δ **Cephei,** is one of the most famous stars in the sky, even though it doesn't look like much. It was the first-ever example of a class of star that now bears its name: the Cepheid variables. As the name suggests, the brightness of δ Cephei varies over the course of a few days (five days and eight hours, to be precise). In this time, its brightness changes by a factor of two. Brighter Cepheid variables take longer to change in brightness.

By measuring how long Cepheid variables take to vary in brightness, astronomers can tell how intrinsically bright they are; they can then compare that brightness with how bright the stars appear from here on Earth to work out how far away they are. Being able to work out such distances is important in astronomy because it's notoriously difficult to say whether the galaxy you're observing is bright because it's close to you or because it's really very bright but is farther away. By looking at Cepheid variable stars in other galaxies, you can work out how far away they are. Astronomers call Cepheid variables *standard candles,* which is what makes δ Cephei so important.

Camelopardalis

Camelopardalis (see Figure 10-7) is one of the faintest of constellations – so faint, in fact, that ancient Greek astronomers considered this part of the sky to be empty, which you may agree with when trying to find Camelopardalis! For that reason, Camelopardalis was quite a late-starter constellation and didn't get its name until the 17th century.

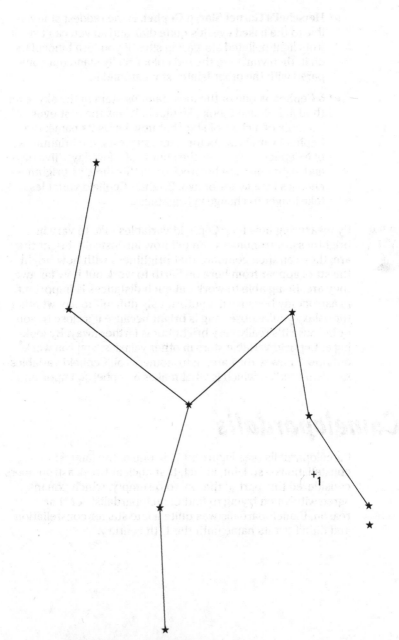

Figure 10-7: Camelopardalis.

Camelopardalis info

Abbreviation: Cam

Genitive: Camelopardalis

Best months for observing in during the evening: Visible all year from

northern latitudes, but highest in the evening from December through May

Camelopardalis has no stars brighter than magnitude 4.

Here are a few things to look for in Camelopardalis.:

Kemble's Cascade (numbered 1 in Figure 10-7) is a run of faint stars that appears to pour like a waterfall into the faint fuzzy open cluster NGC 1502. You should look for Kemble's Cascade using small to medium binoculars (7 x 50s are good) so that you can see all the 20-odd stars that make up the waterfall. Kemble's Cascade is very difficult to find, but it's one of the most beautiful of all binocular objects.

One way to find Kemble's Cascade is to use the first and fifth stars of the W shape of Cassiopeia. Draw a line from *Caph* to *Segin* and go the same distance again. Don't worry if you can't find Kemble's Cascade straight away: it was only discovered by Father Lucian Kemble, using a pair of binoculars, in 1980!

Chapter 11

Stars of December, January and February

. . . .

In This Chapter

▶ Identifying the constellations during the northern hemisphere winter

▶ Spotting the constellations during the southern hemisphere summer

▶ Finding the interesting objects in these constellations

. . . .

During the long dark nights of the winter in the northern hemisphere, you'll be treated to a host of great constellations, including Orion the Hunter with his pair of hunting dogs (Canis Major and Canis Minor) chasing Taurus the Bull across the sky, followed by Gemini the Twins. Some of these constellations are visible from the southern hemisphere, too, during the summer months, and that's where other constellations, such as Carina the Ship and Vela the Sail, rise high in the sky.

Constellations of December, January and February

The constellations in this chapter are best seen in the months of December, January and February, and a map of these is shown in Figure 11-1.

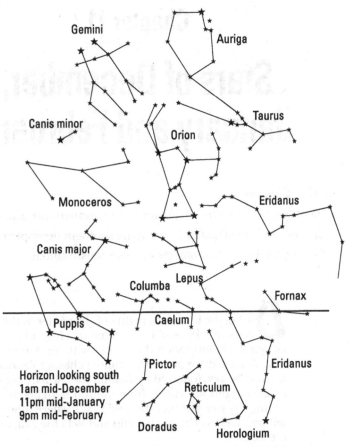

Figure 11-1: The constellations of December, January and February.

This chart shows all the constellations in this chapter. Northern hemisphere stargazers may only be able to see the constellations at the top of this chart, while southern hemisphere stargazers may only be able to see those at the bottom. The horizon line marked in the figure shows what a stargazer at mid-northern latitudes sees.

Remember that as the Earth spins, the position and orientation of the constellations changes. This chart shows their relative positions next to one another. Check monthly star maps for your location to find out where in the sky the constellations actually appear.

Orion

Orion the Hunter (see Figure 11-2) is distinct and easy to spot;
in fact, it's one of the most recognisable constellations in the
whole sky. If you've ever spent any time looking at the night
sky, you may well have seen Orion. He's visible from any point
on the Earth's surface during the months of September to
March, although he's best seen in the evening sky at the end
of this run.

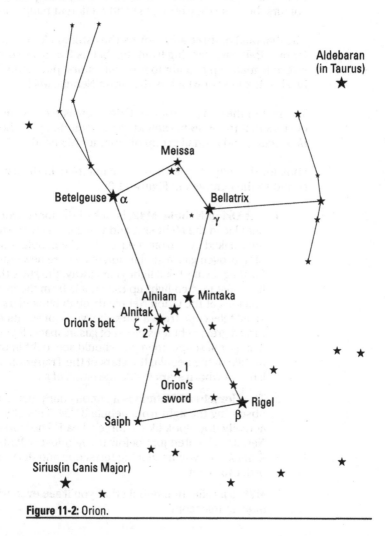

Figure 11-2: Orion.

Orion is made up of four stars in a rectangle shape, with three additional stars in a diagonal across the middle. These stars are all bright enough that you'll see them from a city.

The brightest star in Orion is the blue giant star *Rigel,* in the lower right corner of the rectangle. Almost as bright as *Rigel* is *Betelgeuse,* in the upper left corner. Go and have a look at Orion next time he's up: you should notice that *Betelgeuse* is an orange-red colour, while *Rigel* shines brilliantly blue-white. This colour difference is due to the stars actually being different colours, because they burn at slightly different temperatures.

The diagonal of three stars forms the asterism known as Orion's Belt, and hanging from this belt is Orion's sword. The sword is made up of a number of faint stars, plus some faint fuzzies, the best of which is the Orion Nebula, M42.

The rest of the constellation of Orion consists of a faint curve of stars that make up his shield, arms stretching up above *Betelgeuse,* and a small group of stars for his head.

Look for the following in Orion. The numbers in the list correspond to the numbers in Figure 11-2:

1. The **Orion Nebula, M42,** is one of the most prominent nebulae in the night sky, and you can easily see it with your naked eyes from a dark site. The nebula is a cloud of hydrogen gas, a stellar nursery where new stars are forming. Lying 1,500 light years away, the stars that have just formed light up the nebula from the inside. If you look at the Orion Nebula through binoculars or a small telescope from a dark site, you can see its structure: a grey, nebulous cloud of gas in space. If you're using a telescope, then you should see, right in the middle of the nebula, the stars of the Trapezium asterism – newborn stars inside the cloud of gas.

2. The **Horsehead Nebula** is a famous dark dust cloud obscuring the light from nebula IC434. This object actually does look like a horse's head. The Horsehead Nebula is located just below the leftmost belt star, *Alnitak,* but you need a big telescope and dark skies in order to see it.

3. **M78** is a reflection nebula that you'll see even with a modest telescope.

Orion info

Abbreviation: Ori

Genitive: Orionis

Best months for observing in during the evening: December through February

Name	Bayer Designation	Type	Mag	Distance (ly)
Betelgeuse	α Ori	M2lb	0.58	640
Rigel	β Ori	B9Ia	0.18	773
Bellatrix	γ Ori	B2III	1.64	243
Mintaka	δ Ori	O9II	2.25	916
Alnilam	ε Ori	B0Ia	1.69	1,342
Alnitak	ζ Ori	O9Ib	1.74	817
Na'ir al Saif	ι Ori	O9III	2.75	1,325
Saiph	κ Ori	B0Ia	2.07	721

Orion throughout the year

Orion sits near the ecliptic, or zodiac, line (see Chapter 7), which circles the sky and which the Sun, Moon and planets all move along. As a result, you won't be able to see Orion during a few months of the year when the Sun is very near him in the sky.

Through May, June and July, Orion isn't visible. In the months of April and August, you can catch a glimpse of him only at sunset and sunrise respectively. So the best months for viewing Orion are September through March.

Orion in myth and legend

Orion takes his name from a giant hunter of Greek myth, who in some stories was killed by a scorpion, represented by Scorpius in the night sky. In other stories, Orion was chasing the Seven Sisters (the Pleiades in Taurus) in search of a wife.

Many cultures around the world recognise the stars in Orion as a human figure – in most cases, a man. In Babylon, he was the Shepherd; in ancient Egypt, he was Osiris the Sun God. Islamic astronomers knew him as the Giant, while some Native American tribes knew him as the Winter-maker, because his appearance in the night sky heralds the start of winter.

Canis Major

Canis Major, the Big Dog (see Figure 11-3), is Orion's big hunting dog. This constellation would be a difficult one to spot if it weren't for the really bright star *Sirius,* known as the Dog Star. *Sirius* is the brightest star in the night sky, shining four times brighter than nearby *Rigel,* the brightest star in Orion.

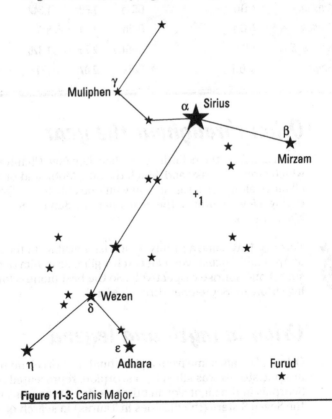

Figure 11-3: Canis Major.

Canis Major info

Abbreviation: CMa

Genitive: Canis Majoris

Best months for observing in during the evening: January through February

Bright Stars List

Name	Bayer Designation	Type	Mag	Distance (ly)
Sirius	α CMa	A0m	−1.46	8.6
Mirzam	β CMa	B1II	1.98	499
Muliphen	γ CMa	B8II	4.11	402
Wezen	δ CMa	F8Ia	1.83	1,791
Adhara	ε CMa	B2II	1.50	431
	η CMa	B5I	2.45	3,196

You may almost be able to make out the stick figure of a dog with two legs, a faint head and a tail, but don't worry if you can't. It's not that obvious!

If you're not sure whether you're looking at *Sirius*, use the stars of Orion's Belt to point to it. Follow a diagonal line down and to the left of Orion as he stands in the sky.

If you're stargazing when *Sirius* is low on the horizon (as it often is from northern Europe in winter), you'll be able to see it twinkle furiously, even looking like it changes colour. Seen from more southerly latitudes, *Sirius* can get very high in the sky indeed and can often appear overhead.

Sirius is bright because it's so close to Earth. It's one of the closest stars to the Sun, just a little under 9 light years away. *Sirius* is the second-closest star to Earth that's visible with the naked eye. The nearest star is α Centauri, or *Rigel Kentaurus*, but that's only visible to observers in the tropics or southern hemisphere.

Using just your eyes, you may be able to make out a faint open cluster in Canis Major, called **M41** (numbered 1 in Figure 11-3), if you've got dark skies and it's high enough above the horizon. You can easily locate it with binoculars or a small telescope. Find *Sirius* and look near it for a faint fuzzy patch of stars with the brightest one being a distinct red colour.

Canis Minor

Orion's second hunting dog is Canis Minor, the Small Dog (see Figure 11-4). Canis Minor doesn't actually look anything like a dog, but it follows Orion across the sky just like Canis Major does, and so the two constellations became Orion's faithful hunting companions.

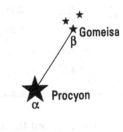

Figure 11-4: Canis Minor.

Like Canis Major, Canis Minor includes a bright star that makes it easy to find. This star is called *Procyon* and lies directly to the left of Orion's Belt, assuming that you're viewing Orion as if he were standing upright in the sky.

Canis Minor info

Abbreviation: CMi

Genitive: Canis Minoris

Best month for observing in during the evening: February

Bright Stars List

Name	Bayer Designation	Type	Mag	Distance (ly)
Procyon	α CMi	F5IV	0.34	11
Gomeisa	β CMi	B8V	2.89	170

The patch of sky around *Procyon* looks more or less empty of stars, meaning that *Procyon* stands out even more.

Auriga

Auriga the Charioteer (see Figure 11-5) represents the pointed helmet of a charioteer. Even more strangely, Auriga is carrying some baby goats – the kids – in his arms. This asterism is in the shape of a narrow wedge of stars near the brightest star in the constellation, *Capella*.

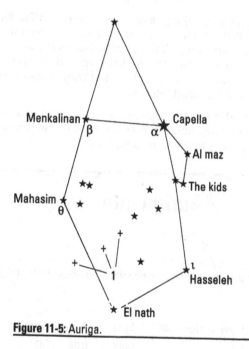

Figure 11-5: Auriga.

You should be able to see that *Capella* is quite a yellow-coloured star, especially when you compare it with the blue star *Rigel* in Orion.

Auriga sits above Orion's head, and so you can easily locate it if you use Orion as a signpost.

Looking out of the galaxy

The Milky Way runs through Auriga, but looking in this direction you're looking directly away from the centre of your galaxy, and so the Milky Way isn't nearly as bright or rich in Auriga as it is elsewhere in the sky.

Look for the following in Auriga. The numbers in the list correspond to the numbers in Figure 11-5:

1. Three open clusters of stars in Auriga, **M36, M37** and **M38,** lie in a line bisecting the line between the stars *Mahasim* and *El Nath.* They're just bright enough that you may be able to see them with your naked eye if you've got very dark skies, but they're certainly visible through a small telescope.

2. *Capella* and **the kids** make a very pretty sight through binoculars, and the yellow colour of the bright star *Capella* should really stand out.

Auriga info

Abbreviation: Aur

Genitive: Aurigae

Best months for observing in during the evening: December through February

Bright Stars List

Name	Bayer Designation	Type	Mag	Distance (ly)
Capella	α Aur	G8III	0.08	42
Menkalinan	β Aur	A2V	1.90	82
Al Maz	ε Aur	F0Ia	3.03	2,037
Mahasim	θ Aur	A0p	2.65	173
Hasseleh	ι Aur	K3II	2.69	512

Gemini

Gemini the Twins (see Figure 11-6) is one of the well-known signs of the zodiac, lying as it does along the ecliptic in the sky (see Chapter 7).

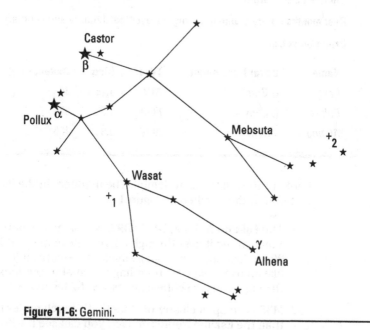

Figure 11-6: Gemini.

The first thing you notice about Gemini is the twin stars *Castor* and *Pollux*, both shining with the same brightness, next to one another in the sky. These two stars form the heads of two stick figures, representing the twins. Can you spot the stick figures? Having the lines drawn on Figure 11-6 makes them look fairly obvious, but just as with all constellation spotting, you may have to convince yourself you can see them!

You can find Gemini using Orion as a signpost. Draw a line from Orion's Belt up through his left shoulder, the red star *Betelgeuse,* and keep on going. The pair of bright stars you come to next are *Castor* and *Pollux.* The bodies and legs of the twins stretch down towards Orion from these brighter stars.

Gemini info

Abbreviation: Gem

Genitive: Geminorum

Best months for observing in during the evening: January and February

Bright Stars List

Name	Bayer Designation	Type	Mag	Distance (ly)
Castor	α Gem	A2V	1.16	34
Pollux	β Gem	K0III	1.58	52
Alhena	γ Gem	A0IV	1.93	105

Look for the following in Gemini. The numbers in the list correspond to the numbers in Figure11-6:

1. The **Eskimo Nebula, NGC 2392,** is a planetary nebula. You can see it best through a large telescope of at least 20cm diameter. Through a smaller telescope, it looks like an oval shape surrounding a faint star, but looking through a larger scope enables you to see more detail.

2. **M35** is an open cluster of stars that's much easier to find than the Eskimo Nebula; in fact, you can see it with just your eyes if you've got a good dark site. If you want to see M35 in more detail, use binoculars or a low-power telescope.

Taurus

Taurus the Bull (see Figure 11-7) is another of the signs of the zodiac, sitting on the ecliptic (see Chapter 7). It's a large and rich constellation, easy to spot due to the bright red star *Aldebaran,* the red eye of the Bull.

Aldebaran lies within a large and diffuse open cluster called the Hyades, a V shape of stars with *Aldebaran* at one tip. The horns of Taurus stretch from this V shape towards the constellation Gemini.

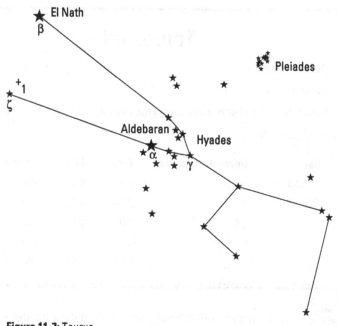

Figure 11-7: Taurus.

Like with so many other constellations, you can find Taurus using the stars of Orion as signposts. If you follow the line of the three stars of Orion's Belt up and to the right of Orion through his shield, you'll find *Aldebaran* and Taurus.

Look for the following in Taurus. The numbers in the list correspond to the numbers in Figure 11-7:

1. The **Crab Nebula, M1,** is a supernova remnant, a shell of gas around a dead star that exploded almost 1,000 years ago. It's very faint, at the limit of your eyesight under darks skies. Through binoculars you can see only a small faint fuzzy patch, but through a medium-sized telescope you can start to see a bit of detail.

2. The **Pleiades, M45,** is an open cluster of stars also known as the Seven Sisters because, with good eye-sight, you can make out seven of them. The Pleiades look beautiful through binoculars, showing dozens – hundreds – of faint stars.

Taurus info

Abbreviation: Tau

Genitive: Tauri

Best months for observing in during the evening: December and January

Bright Stars List

Name	Bayer Designation	Type	Mag	Distance (ly)
Aldebaran	α Tau	K5III	0.87	65
El Nath	β Tau	B7III	1.65	131
	γ Tau	B7III	2.85	368
	ζ Tau	B4III	2.97	417
Alcyone	η Tau	B7III	2.85	368

Test your eyesight: how many stars in the Pleiades can you count?

Lepus

Lepus the Hare (see Figure 11-8) lies beneath Orion's feet and is a dim and relatively indistinct constellation. Its brightest stars are *Arneb* and *Nihal*. *Arneb* is a giant star that may one day go supernova, brightening considerably.

Lepus is easy to find, because you can, once again, use Orion as a signpost. Just follow a line down from his belt between the bright stars *Saiph* and *Rigel,* and you get to Lepus. However, for stargazers in far northern latitudes, Lepus never gets very high in the sky, which makes it quite a tricky target.

Look for **M59,** a small but beautiful globular cluster. M59 (numbered 1 in Figure 11-8) looks good through a medium telescope under dark skies. It appears as a fuzzy patch if you try to find it with binoculars.

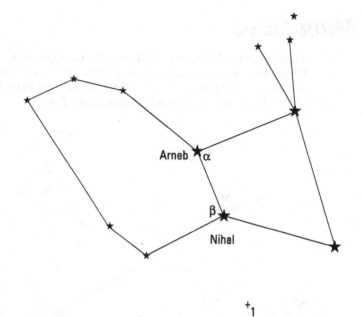

Figure 11-8: Lepus.

Lepus info

Abbreviation: Lep

Genitive: Leporis

Best month for observing in during the evening: January

Bright Stars List

Name	Bayer Designation	Type	Mag	Distance (ly)
Arneb	α Lep	FOIb	2.58	1,283
Nihal	β Lep	G5II	2.81	159

Monoceros

Monoceros the Unicorn (see Figure 11-9) lies between Canis Minor and Canis Major, making it quite easy to locate, although you may be hard-pressed to find any bright stars in that part of the sky; Monoceros is quite a dim constellation.

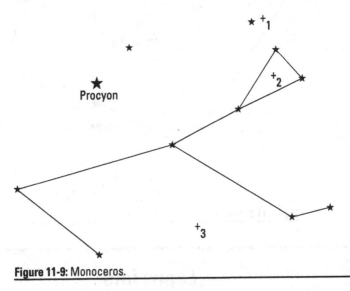

Figure 11-9: Monoceros.

Draw a line between the two bright stars of *Sirius* and *Procyon;* halfway along that line is Monoceros.

Despite the absence of any really bright stars in Monoceros, you can find a few great faint fuzzies in this constellation.

Look for the following in Monoceros. The numbers in the list correspond to the numbers in Figure11-9:

1. The **Christmas Tree Nebula, NGC 2264,** is just about visible to your naked eye under dark skies, but it'll really stand out if you look at it through binoculars. It's got a distinct Christmas tree shape, and a lot of interesting features are visible through a big telescope. The most obvious is the Cone Nebula, a dark cone shape jutting into the Christmas tree.

Monoceros info

Abbreviation: Mon

Genitive: Monocerotis

Best months for observing in during the evening: January and February

Monoceros has no stars brighter than magnitude 3.9.

2. The **Rosette Nebula, NGC 2237,** is a large cloud of hydrogen gas in space, lit up by the stars in and around it. From a dark site, you may be able to make out these stars with binoculars or a telescope, but in order to see the cloud of gas around the stars, you'll need a big telescope with low magnification.

3. **M50,** an open cluster of stars, looks very pretty through binoculars and small telescopes and is relatively easy to find, thanks to two bright stars nearby. Draw a line from *Sirius* to *Procyon;* about a third of the way along that line and offset slightly to the right you'll find M50.

Puppis

Puppis the Poop-deck (see Figure 11-10) or Stern, is the third part of the old constellation Argo Navis, the other two being Carina the Keel and Vela the Sail.

Puppis covers a large area, and so it may be a little tricky to find. If you draw a line from *Canopus* to *Sirius,* the two brightest stars in the sky, Puppis lies to the left of that line as you move from *Canopus* to *Sirius.*

The Milky Way runs through Puppis, as it does through Carina and Vela, making it a great constellation for faint fuzzies. You can see open clusters a-plenty through binoculars.

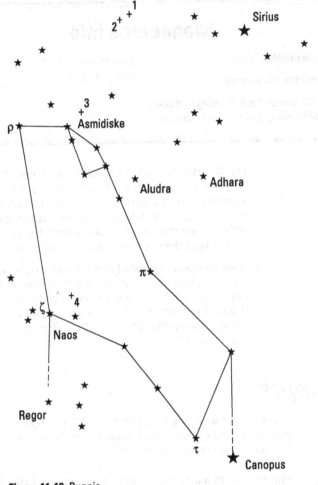

Figure 11-10: Puppis.

Look for the following in Puppis. The numbers in the list correspond to the numbers in Figure 11-10:

1. **M47** is an open cluster that you can see with your eyes if the skies are dark enough. Through a small telescope, M47 looks far coarser than the richer, sparkling open cluster M46 next door.

Puppis info

Abbreviation: Pup

Genitive: Puppis

Best months for observing in during the evening: January and February

Bright Stars List

Name	Bayer Designation	Type	Distance (ly)
Naos	ζ Pup	O5ia	1,399

2. **M46** is another open cluster, larger and richer than M47, with more faint stars, which you can get a good view of through a small telescope with a low-power eyepiece attached.

3. **M93** is another open cluster, smaller and brighter than M46, with a distinct triangular shape to it.

4. **NGC 2477** is one of the best open clusters in the sky. Looking at NGC 2477 through a small telescope, it appears almost like a globular cluster, with a very spherical shape, but without the dense core of a globular cluster. The stars in NGC 2477 stand out quite clearly as individuals, because NGC 2477 is much closer than globular clusters tend to be.

Caelum

Caelum the Chisel (see Figure 11-11) is a tiny constellation, with no bright stars in it, lying between Eridanus and Columba. Caelum is a southern sky constellation, and so European astronomers named it only in the 18th century. Caelum is meant to represent a chisel, but even its brightest stars can get lost in the glare of bright city lights. Under darker skies, you may be able to populate this blank patch of sky with a few stars and join the dots to make a chisel. Good luck!

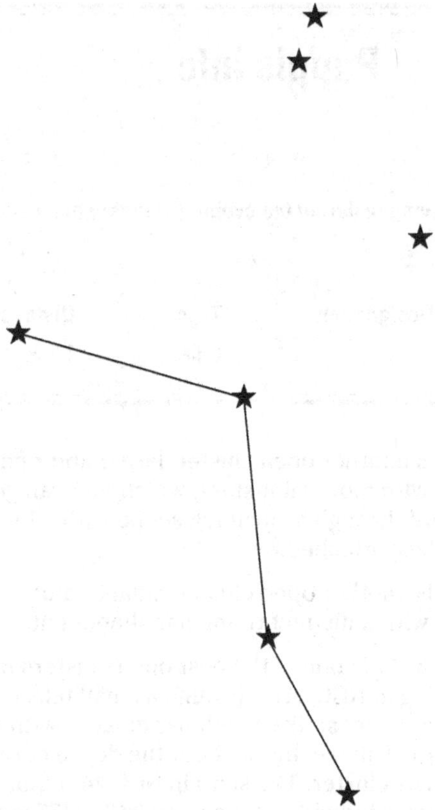

Figure 11-11: Caelum.

Finding Caelum is tricky, but once again Orion comes to the rescue. If you draw a line from *Betelgeuse* through Orion's Belt and his sword, then through the constellation Lepus, and keep going the same distance again, you've found Caelum.

Caelum info

Abbreviation: Cae

Genitive: Caeli

Best months for observing in during the evening: December and January

Caelum has no stars brighter than magnitude 4.

Caelum has no bright faint fuzzies, and the ones that are there appear very faint and small; you need a big telescope to even have a chance of tracking them down.

Columba

Columba the Dove (see Figure 11-12) lies next to Caelum, and so it's best seen from southern skies. The two brightest stars in Columba, *Phact* and *Wazn,* are bright enough to be seen from a town, but the fainter stars will vanish under the glare of light pollution.

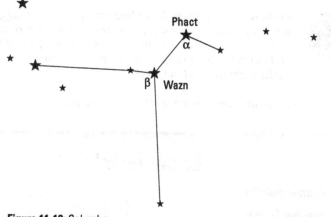

Figure 11-12: Columba.

Columba info

Abbreviation: Col

Genitive: Columbae

Best month for observing in during the evening: January

Bright Stars List

Name	Bayer Designation	Type	Mag	Distance (ly)
Phact	α Col	B7IV	−2.65	268
Wazn	β Col	K1III	3.12	86

Follow the line of Orion's sword down through Lepus, and go the same distance again to find Columba.

Like Caelum, Columba is inconspicuous and doesn't really have any faint fuzzies worth tracking down.

Eridanus

Eridanus the River (see Figure 11-13) is a huge constellation, winding its way through the sky from beside the star of Orion's right foot, *Rigel,* down to the bright star *Achernar* in the far southern skies.

Eridanus is so long and winding that you may struggle to track it through the sky, especially because of its lack of bright stars; the only really bright one is *Achernar.* To find the other end of Eridanus, find the star *Cursa,* which sits just above and to the right of *Rigel* in Orion.

Eridanus contains no distinct faint fuzzies.

Eridanus info

Abbreviation: Eri

Genitive: Eridani

Best months for observing in during the evening: November through January

Bright Stars List

Name	Bayer Designation	Type	Mag	Distance (ly)
Achernar	α Eri	B3V	0.45	144
Cursa	β Eri	A3III	2.78	89
Zaurak	γ Eri	M1IIIb	2.97	221
Acamar	θ Eri	A4III	2.88	161

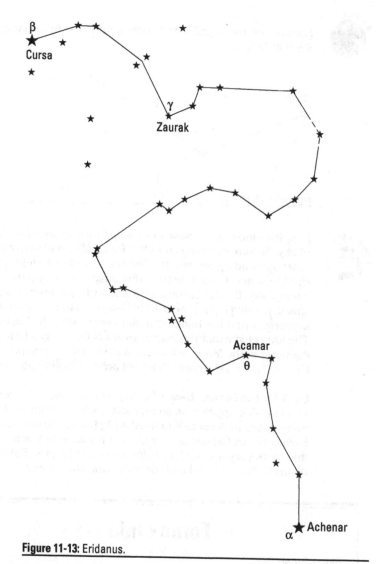

Figure 11-13: Eridanus.

Fornax

Fornax the Furnace (see Figure 11-14) is another faint, unimpressive southern hemisphere constellation nestled in a bend of the constellation Eridanus, the celestial river.

Fornax lies about half way between *Achernar* in Eridanus and *Rigel* in Orion.

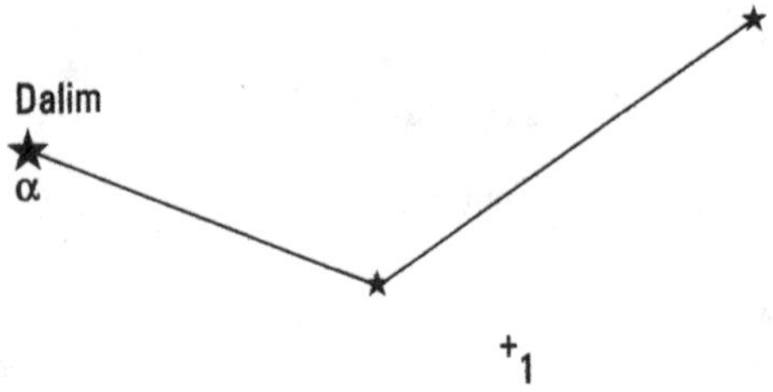

Dalim
★
α

+
1

Figure 11-14: Fornax.

Despite – indeed, because of – its appearance as a blank patch of sky, Fornax was chosen as the focus of one of astronomy's most astounding images: the Hubble Ultra-Deep Field. The Hubble Space Telescope took this image of a tiny patch of 'empty' sky. It's the farthest astronomers have ever looked into space, peering back almost 13 billion years to the very early universe, just a few hundred million years after the Big Bang. The area of sky that the image covers is tiny, only a tenth of the diameter of the Moon in the sky, and in that 'empty' patch, the Hubble Space Telescope observed around 10,000 galaxies!

Look for the **Fornax Dwarf Galaxy**, (numbered 1 in Figure 11-14) which appears as an incredibly dim and diffuse faint fuzzy patch in Fornax. You need a big telescope to see the Fornax Dwarf Galaxy, and even then it's hard to track down due to its very low surface brightness. Finding the Fornax Dwarf Galaxy is a real test for telescope stargazers!

Fornax info

Abbreviation: For

Genitive: Fornacis

Best months for observing in during the evening: November and December

Bright Stars List

Name	Bayer Designation	Type	Mag	Distance (ly)
Dalim	α For	F8V	3.80	46

Horologium

Horologium the Clock (see Figure 11-15) is another faint southern hemisphere constellation.

You can find Horologium sitting on the banks of Eridanus the River, near the bright star *Achernar*.

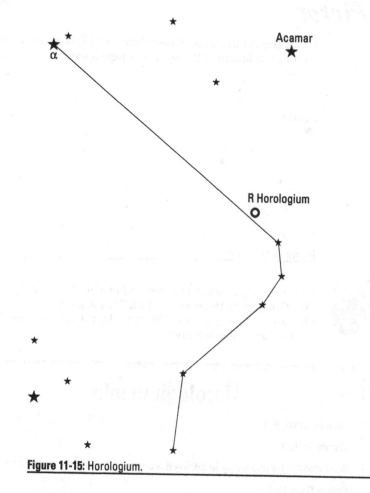

Figure 11-15: Horologium.

You won't find a lot to look at in Horologium in terms of faint fuzzies, but it does have one interesting star in it: *R Horologium,* a variable star, meaning that it changes its brightness. Lots of stars are variables, but none vary quite so much as *R Horologium*. At its brightest, *R Horologium* is the second brightest star in the constellation. *R Horologium* then fades away to invisibility, only to brighten up again and repeat the cycle. It takes around 400 days to go from bright through dim through bright again.

Pictor

Pictor the Painter's Easel (see Figure 11-16) is another small, faint constellation in the southern hemisphere.

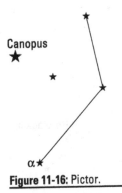

Canopus

α

Figure 11-16: Pictor.

Pictor lies right next to the second brightest star in the sky, *Canopus,* in the constellation Carina, so finding Pictor shouldn't be too difficult. After you do, though, you won't see a lot to keep your attention.

Horologium info

Abbreviation: Hor

Genitive: Horologii

Best months for observing in during the evening: November and December

Bright Stars List

Name	Bayer Designation	Type	Mag	Distance (ly)
	α Hor	K1III	3.85	117

Pictor info

Abbreviation: Pic

Genitive: Pictoris

Best months for observing in during the evening: December through February

Bright Stars List

Name	Bayer Designation	Type	Mag	Distance (ly)
	α Pic	A7IV	3.24	99

Reticulum

Reticulum the Reticle (see Figure 11-17) is a small and faint constellation.

You can find Reticulum between the equally faint Horologium constellation and the Large Magellanic Cloud (see Chapter 15). You won't see many bright stars here and even fewer good faint fuzzies.

★

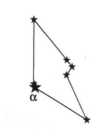

Figure 11-17: Reticulum.

Reticulum info

Abbreviation: Ret

Genitive: Reticuli

Best months for observing in during the evening: February and March

Bright Stars List

Name	Bayer Designation	Type	Mag	Distance (ly)
	α Ret	G7III	3.33	163

Chapter 12

Stars of March, April and May

*W*hile not as dark as the winter months, the shoulder seasons of spring and autumn still give you a great chance to see lots of interesting objects in the night sky. In the months of March, April and May, a parade of animals are on display, including zodiac constellations like Leo the Lion and Cancer the Crab. Northern skies are illuminated by Boötes the Herdsman, marked out by the bright star *Arcturus*, the fourth brightest in the night sky, while in southern skies you see the mighty Centaurus, half man, half horse, whose front foot is α Centauri, also known as *Rigel Kentaurus*, the closest bright star to the Sun and the third brightest in the night sky.

Constellations of March, April and May

The constellations in this chapter are best seen in the months of March, April and May, and a map of these is shown in Figure 12-1.

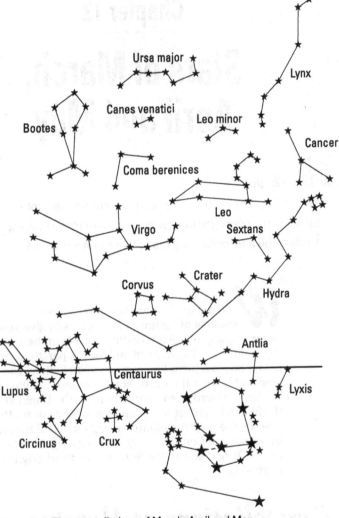

Figure 12-1: The constellations of March, April and May.

This chart shows all the constellations in this chapter. Northern hemisphere stargazers may only be able to see the constellations at the top of this chart, while southern hemisphere stargazers may only be able to see those at the bottom. The horizon line marked in the figure shows what a stargazer at mid-northern latitudes will see.

Remember that, as the Earth spins, the position and orientation of the constellations changes. Figure 12-1 shows the relative positions of the constellations next to one another. Check monthly star maps for your location to find out where in the sky the constellations actually appear.

Boötes

Boötes (pronounced 'boh-oh-teez') the Herdsman (see Figure 12-2) is roughly kite-shaped, and is herding the great bear, Ursa Major. (For more on Ursa Major, see Chapter 10.)

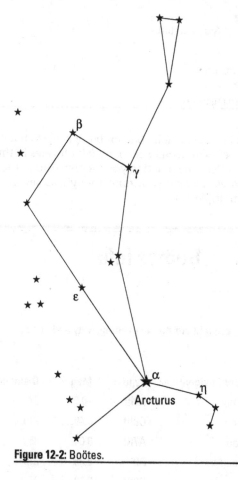

Figure 12-2: Boötes.

By far the brightest star in Boötes is *Arcturus,* meaning bear-guard. This giant star is relatively close to Earth, so it's very bright in the night sky. *Arcturus* has a warm orange colour.

As well as having the bright orange star *Arcturus* to guide you to Boötes, a well-known mnemonic can help you track him down. Using the bright stars of the Plough in Ursa Minor, you can follow the curve of the Plough's handle and 'arc to *Arcturus*', as shown in Figure 12-3.

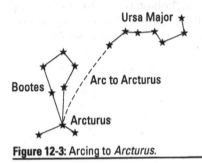

Figure 12-3: Arcing to *Arcturus.*

Boötes sits in an especially barren bit of sky when it comes to faint fuzzies. You won't see many faint fuzzies in this area, even with a telescope. Casting your scope around the brighter stars in Boötes is worth your time, though, as several are double or multiple stars.

Boötes info

Abbreviation: Boo

Genitive: Boötis

Best months for observing in during the evening: May and June

Bright Stars List

Name	Bayer Designation	Type	Mag	Distance (ly)
Arcturus	α Boo	K2III	−0.05	37
Nekkar	β Boo	G8III	3.49	219
Seginus	γ Boo	A7III	3.04	85
Izar	ε Boo	A0	2.35	210
Muphrid	η Boo	G0IV	2.68	37

Centaurus

Centaurus the Centaur (see Figure 12-4) is a huge southern hemisphere constellation, meant to represent the mythical centaur – half man, half horse. After the dots are joined, you may be able to convince yourself that you're seeing a centaur. Your eyes may be drawn away, though, by the two brilliantly bright stars of the centaur's feet α and β Centauri, the third and eleventh brightest stars in the night sky, known as the southern pointers.

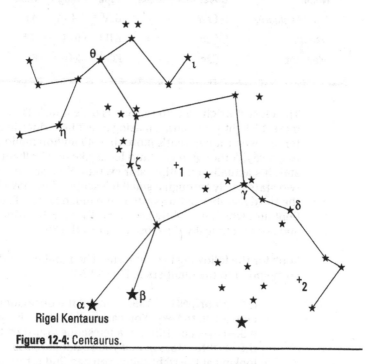

Figure 12-4: Centaurus.

No southern pole star exists, and so stargazers and navigators in the southern hemisphere have a bit of a harder time finding south than their northern cousins have finding north. To find south, you have to use the constellation Crux, the Southern Cross, along with α and β Centauri, the southern pointers. Draw a line down from the long axis of the Southern Cross, and simultaneously draw a line that cuts α and β Centauri apart. Where those two lines meet is the southern pole. See Figure 2-3 in Chapter 2 for a diagram.

Centaurus info

Abbreviation: Cen

Genitive: Centauri

Best months for observing in during the evening: April and May

Bright Stars List

Name	Bayer Designation	Type	Mag	Distance (ly)
Rigel Kentaurus	α Cen	G2V	−0.01	4.6
Hadar	β Cen	B1III	0.61	525
Menkent	θ Cen	K0III	2.06	61

The closest bright star to the Sun is α Centauri. This star is a mere 4.6 light years away, making it just next door in galactic terms, even though that's more than 40 million million kilometres away! Although α Centauri looks like one brilliantly bright star, it's actually a multiple star system. You can see the main two stars easily through a small telescope. You may also find the third member of this system, the much fainter Proxima Centauri, which is slightly nearer Earth than the brighter pair, and so is technically the closest star to the Sun.

Look for the following in Centaurus. The numbers in the list correspond to the numbers in Figure 12-4:

1. ω **Centauri, NGC 5139,** is one of the best globular clusters in the sky. You can easily see it with your naked eye, and through a telescope you start to see the brighter stars in it. If you're not sure that you're looking at the right thing, you can find ω Centauri by drawing a line north from α and β Crucis (also called Acrux and Mimosa, respectively).

2. The **Blue Planetary Nebula, NGC 3918,** looks like a small blue planet, hence the name! You need a telescope to see the Blue Planetary Nebula properly. Use a low magnification to begin with until you're sure that you've found it, and then change eyepieces to increase the magnification until you can see the blue disk.

Carina

Carina the Keel (see Figure 12-5) is part of a huge old constellation called Argo Navis, the Ship, which, due to its unwieldy size, was split in three:

- Carina the Keel
- Vela the Sail
- Puppis the Stern

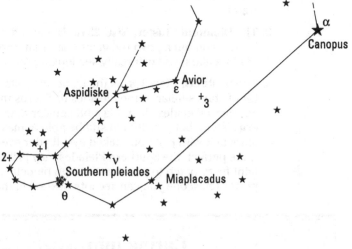

Figure 12-5: Carina.

You can't see Carina from far northern latitudes, which is a pity for northern stargazers, because it's got lots going for it.

Southern stargazers can make sure that they're looking at Carina by using Orion and *Sirius* as signposts. Draw a line from Orion's Belt to *Sirius,* and then turn left down the length of Canis Major towards *Canopus,* the brightest star in Carina.

Canopus is the second brightest star in the night sky, outshone only by *Sirius* in Canis Major. If you're stargazing from southern latitudes, you can often see these two stars in the sky together, *Sirius* shining about twice as brightly as *Canopus.*

Look for the following in Carina. The numbers in the list correspond to the numbers in Figure 12-5:

1. The **Eta Carinae Nebula, NGC 3372,** is one of the brightest naked-eye faint fuzzies in the sky, brighter even than the Orion Nebula, but it's visible high in the sky only from southern latitudes. It's a cloud of gas surrounding a bright hypergiant star Eta Carinae. Try looking at it through binoculars for an even more spectacular view.

2. The **Wishing Well Cluster, NGC 3532,** is an open cluster of beautiful stars near the Eta Carinae Nebula. It looks great through a telescope, like coins twinkling in a well of water.

3. The **Diamond Cluster, NGC 2516,** is a naked-eye open cluster containing two red giant stars that, through a pair of binoculars, you can really see looking red.

4. The **Southern Pleiades,** as the name suggests, is a cluster of stars similar to the Pleiades in Taurus the Bull. The Southern Pleiades cluster is a bit smaller than its northern sisters, but you should still be able to make out some stars in it with just your naked eye. If your eyesight isn't quite perfect, the Southern Pleiades may just look like a faint fuzzy patch to you, but through binoculars or a low-power telescope you can see lots of stars in this cluster.

Carina info

Abbreviation: Car

Genitive: Carinae

Best months for observing in during the evening: January through April

Bright Stars List

Name	Bayer Designation	Type	Mag	Distance (ly)
Canopus	α Car	F0Ib	−0.72	313
Miaplacadus	β Car	A2IV	1.67	111
Avior	ε Car	K3III	1.86	632
Aspidiske	ι Car	A8Ib	2.21	692

Vela

Vela is the Sail (see Figure 12-6) of the ship Argo Navis, the other two parts being Carina the Keel and Puppis the Stern. Like the other constellations that make up the Ship, Vela is big. As befits a sail, it sits above Carina and Puppis when those two are pictured as a boat.

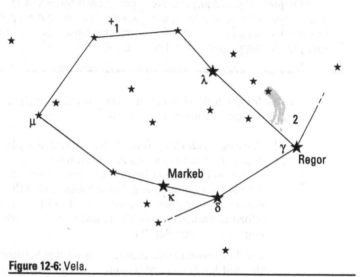

Figure 12-6: Vela.

Vela lies in the part of the sky roughly between the very bright star *Canopus* and the constellations of Crux and Centaurus, making it quite easy to find for southern stargazers.

The brightest star in Vela is γ Velorum, better known as *Regor*. This name was coined in 1960 by American astronaut Gus Grissom in honour of his friend and fellow astronaut Roger Chaffee. *Regor* is Roger spelled backwards! If you have a high-power pair of binoculars steadied on a tripod, you may be able to make out that *Regor* is actually a binary star – two stars side by side. To your naked eye, it looks like one star, but zoom in with binoculars and you can see both.

Like Carina and Puppis, the Milky Way runs through Vela, making it a great constellation for faint fuzzies, particularly planetary nebulae.

The False Cross

Two stars in Vela make up half of the False Cross, the bane of celestial navigators in the southern hemisphere. The Southern Cross (see Chapter 15) points towards the south celestial pole and therefore lets you find due south . . . assuming that you use the correct cross. The False Cross looks very similar to the Southern Cross and sits very near the real thing, and so it may lead you astray. You can tell the difference if you know what to look for, but new stargazers may get tripped up the first few times they try to find due south!

Look for the following in Vela. The numbers in the list correspond to the numbers in Figure 12-6:

1. The **Southern Ring Nebula, NGC 3132,** is a planetary nebula that looks like its northern namesake in the constellation Lyra (see Chapter 13). You can make out this nebula using binoculars, but it looks better through a telescope. Because it's quite a bright object, you can afford to go up to quite high magnifications, which allows you to see some detail in it.

2. The **Vela Supernova Remnant** is a huge feature in the sky, but it's very hard to make out because it's so dim. Because it's so big, use large binoculars on a tripod; even then, you may struggle to find anything but the dimmest hint of fuzziness.

Vela info

Abbreviation: Vel

Genitive: Velorum

Best months for observing in during the evening: February through April

Bright Stars List

Name	Bayer Designation	Type	Mag	Distance (ly)
Regor	γ Vel	O9I	1.75	840

Cancer

Cancer the Crab (see Figure 12-7) is one of the constellations that lie along the zodiac, or ecliptic. It has no bright stars in it, and so finding Cancer can be quite difficult.

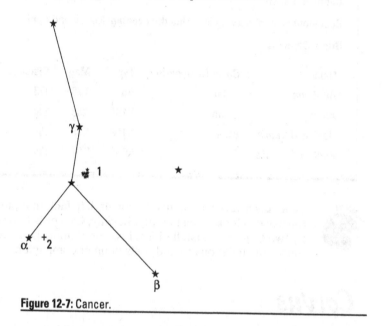

Figure 12-7: Cancer.

Because Cancer sits on the ecliptic, you can often find planets or the Moon somewhere in this constellation. The Sun passes through Cancer between 20 July and 10 August each year.

Look for the following in Cancer. The numbers in the list correspond to the numbers in Figure 12-7:

1. The **Beehive Cluster, M44,** is a great open cluster. You can catch it with the naked eye under dark skies, but binoculars show it best as a faint group of stars in a vaguely triangular shape, like an old-fashioned beehive. The bright-ish stars γ and δ Cancri lie just above it.

2. **M67** is another open cluster, much fainter than M44, but still visible with binoculars or a telescope. M67 is right next to α Cancri, which makes finding it a bit easier, because you can use this star as a marker.

Cancer info

Abbreviation: Can

Genitive: Cancri

Best months for observing in during the evening: March and April

Bright Stars List

Name	Bayer Designation	Type	Mag	Distance (ly)
Acubens	α Can	A5	4.26	173
Al Tarf	β Can	K4III	3.53	290
Asellus Borealis	γ Can	A1IV	4.66	158
Asellus Australis	δ Can	K0III	3.94	136

Cancer is a very faint constellation and, in fact, that part of the sky can look almost empty. Luckily, Cancer is book-ended by two brighter constellations, Leo and Gemini. Look between these constellations to find their dimmer companion.

Corvus

Corvus the Crow (see Figure 12-8) is a small but easily recognisable constellation. Its four brighter stars make a distinct shape, much like the Keystone asterism in the constellation Hercules. Corvus is paired with the nearby constellation of Crater the Cup. Legend has it that Apollo sent Corvus to fetch water in the cup.

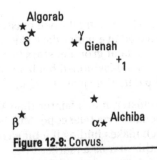

Figure 12-8: Corvus.

Corvus info

Abbreviation: Crv

Genitive: Corvi

Best months for observing in during the evening: April and May

Bright Stars List

Name	Bayer Designation	Type	Mag	Distance (ly)
Alchiba	α Crv	F0IV	4.02	48
Kraz	β Crv	G5II	2.65	140
Gienah	γ Crv	B8III	2.58	165
Algorab	δ Crv	B9V	2.94	88

Look for the **Antennae Galaxies, NGC 4038** and **4039,** in Corvus. They're numbered 1 in Figure 12-8. You can only see these two colliding galaxies through a medium or large telescope.

Crater

Crater the Cup (see Figure 12-9) is next to Corvus the Crow, but its stars are much fainter, making the constellation harder to see.

In Crater, look for **NGC 3511** and **3513,** two galaxies close enough together that you can see them both in the field of view of a telescope. They're labelled 1 in Figure 12-9. They're not especially bright, so you need a larger telescope to see them well.

Crater info

Abbreviation: Crt

Genitive: Craeteris

Best months for observing in during the evening: April and May

Bright Stars List

Name	Bayer Designation	Type	Mag	Distance (ly)
Alkes	α Crt	K1III	4.08	174
	δ Crt	K0III	3.56	195

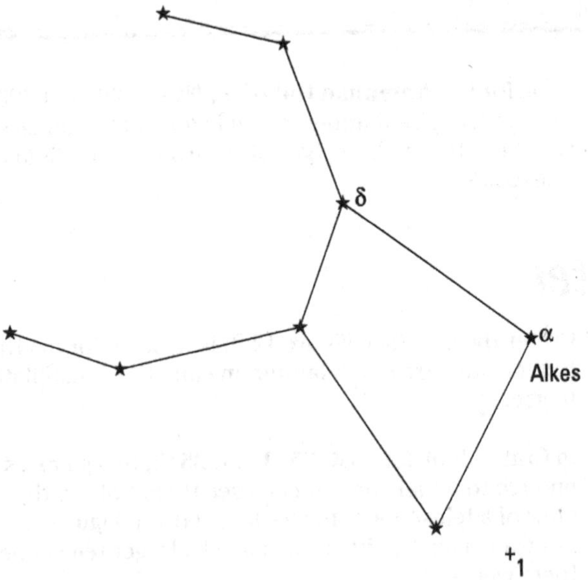

Figure 12-9: Crater.

Leo

Leo the Lion (see Figure 12-10) is one of the brightest and best of the zodiac constellations, along with Taurus, Gemini,

Scorpius and Sagittarius. The head of the lion is made up of the Sickle asterism, at the base of which is the bright star *Regulus*.

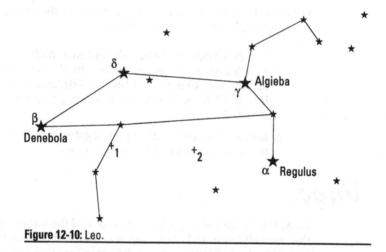

Figure 12-10: Leo.

The backwards question-mark shape that forms the head of Leo the Lion is known as the Sickle, and is one of the more famous asterisms in the night sky.

Leo info

Abbreviation: Leo

Genitive: Leonis

Best months for observing in during the evening: April and May

Bright Stars List

Name	Bayer Designation	Type	Mag	Distance (ly)
Regulus	α Leo	B7V	-0.52	77
Denebola	β Leo	A3V	2.14	36
Algieba	γ Leo	K0III	2.01	126
Zosma	δ Leo	A4V	2.56	58

Leo sits in the ecliptic, and so the planets, Moon and Sun pass through this constellation. The Sun is in Leo between 10 August and 16 September.

Look for the following in Leo The numbers in the list correspond to the numbers in Figure12-10:

1. The **Leo Triplet – M65, M66 and NGC 3628 –** is three galaxies all visible in the same field of view. You can just manage to make out the Leo Triplet using binoculars, but a larger-aperture telescope lets you see more detail.

2. **M95** and **M96** is another pair of galaxies, just some of the many other bright galaxies in Leo.

Virgo

Virgo the Virgin (see Figure 12-11) is one of the largest of all the constellations and contains many interesting objects, including the brilliantly bright white star *Spica* and a nearby cluster of galaxies.

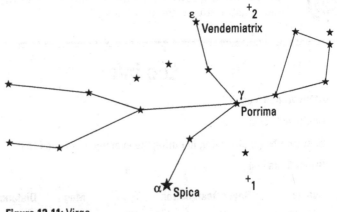

Figure 12-11: Virgo.

 You can continue the route you take to find *Arcturus* in Boötes (see the 'Boötes' section, earlier in this chapter) to find *Spica* in Virgo. Using the handle of the Plough in Ursa Major, you arc to *Arcturus* and then drive a spike to *Spica*, which makes finding the otherwise innocuous constellation of Virgo much easier.

Virgo info

Abbreviation: Vir

Genitive: Virginis

Best months for observing in during the evening: May and June

Bright Stars List

Name	Bayer Designation	Type	Mag	Distance (ly)
Spica	α Vir	B1V	0.98	262
Porrima	γ Vir	F0V	2.74	39
Vindemiatrix	ε Vir	G8III	2.85	102

Spica represents a sheaf of wheat held by Virgo; when it begins to appear in the evening sky after sunset, it signals that (northern hemisphere) spring is on the way.

If you look in the direction of Virgo, you're looking towards a cluster containing between 1,000 and 2,000 galaxies spread over millions of light years. They're all pretty far away – 20 times farther than the bright Andromeda Galaxy, M31 – so they're much fainter, but you can still see many of them through a telescope.

Look for the following in Virgo. The numbers in the list correspond to the numbers in Figure 12-11:

1. The **Sombrero Galaxy, M104,** is one of the brighter galaxies in Virgo, but it's still tricky to find. You can locate the Sombrero Galaxy by using large binoculars or a small telescope. This galaxy forms the right-angled corner of a triangle in which the other two points are the bright stars α and γ Virginis.

2. The **Virgo Cluster** contains many galaxies visible through a telescope, but they're all rather fuzzy and faint, which means you have to take care to find them. Many of these galaxies may lie in the same field of view of your telescope, so you may see two or more at once. The Virgo Cluster actually extends north into the constellation of Coma Berenices.

The star *Porrima* is a nice double star, and shows up well through binoculars or a small telescope.

Antlia

Antlia the Water Pump (see Figure 12-12) is a small, faint constellation of the southern skies. It lies north of Centaurus. Vela the Sail and the bright stars of γ and λ Velorum point towards it.

Figure 12-12: Antlia.

You won't see a lot in this dim patch of sky.

Canes Venatici

Canes Venatici (see Figure 12-13) are the Hunting Dogs of Boötes the Herdsman. They lie between Boötes and Ursa Major, the bear Boötes is herding, beneath the curve of the handle of the Plough.

Antlia info

Abbreviation: Ant

Genitive: Antlae

Best months for observing in during the evening: March and April

No stars are brighter than magnitude 4.

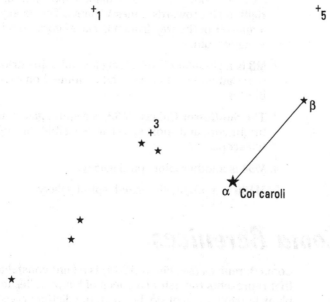

Figure 12-13: Canes Venatici.

Look for the following in Canes Venatici. The numbers in the list correspond to the numbers in Figure 12-13:

1. The **Whirlpool Galaxy, M51,** is one of the most famous galaxies in the sky. It lies face on to Earth and was the first galaxy seen as a spiral. This galaxy is actually much closer to Ursa Major than it is to the bright-ish stars

of Canes Venatici, and you can find it by using the two end stars of the handle of the Plough. Draw a line from ξ Ursae Majoris to η Ursae Majoris and then turn left at right angles towards Canes Venatici. The galaxy is about a quarter of the way from η Ursae Majoris to β Canum Venaticorum.

2. **M3** is a globular cluster, lying just off a line drawn from β to α Canum Venaticorum and continued on to *Arcturus* in Boötes.

3. **The Sunflower Galaxy, M63,** is a spiral galaxy with a bright core and short spiral arms visible through a large telescope.

4. **M94** is another faint spiral galaxy.

5. **M106** is a bright, elongated, spiral galaxy.

Coma Berenices

Coma Berenices (see Figure 12-14) is a faint constellation that represents the hair of a queen of Egypt. It lies in the blank-looking patch of sky between the distinct constellations Boötes and Leo, which makes locating it a bit easier.

Look for the following in Coma Berenices. The numbers in the list correspond to the numbers in Figure 12-14:

1. **M53** is a globular cluster right next to α Comae Berenices, a faintish star that lies about half way between β Leo and *Arcturus,* which makes it quite easy to find.

Canes Venatici info

Abbreviation: CVn

Genitive: Canum Venaticorum

Best months for observing in during the evening: May

Bright Stars List

Name	Bayer Designation	Type	Mag	Distance (ly)
Cor Coroli	α CVn	A0	2.89	110

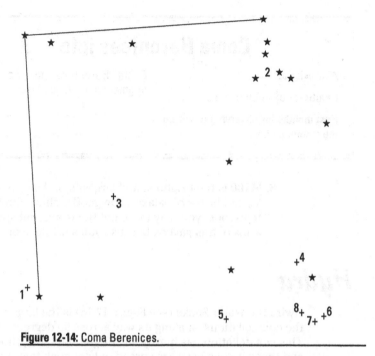

Figure 12-14: Coma Berenices.

2. The **Coma Star Cluster** is the hair of the queen. It falls in cascades from the star γ Comae Berenices. With just your eyes, you can tell that something is different about this patch of sky. The Coma Star Cluster is actually a star cluster around 250 light years away, so close that the stars appear spread out over quite a large area of sky compared with the more densely packed Pleiades, M45, for example. If you turn your binoculars on the Coma Star Cluster, you're in for a treat, because stars fill your field of view.

3. **The Black-Eye Galaxy, M64,** looks striking through large telescopes, with a dark band of dust giving it the appearance of a black eye.

4. **M85** is a distant, faint lenticular (lens-shaped) galaxy.

5. **M88 and M91** are two galaxies that share the same field of view, both spiral galaxies.

6. **M98** is another faint spiral galaxy.

7. **M99** is yet another faint spiral galaxy.

Coma Berenices info

Abbreviation: Com

Genitive: Comae Berenices

Best months for observing in during the evening: May

Coma Berenices has no stars brighter than magnitude 4.

---

8. **M100** is the brightest and largest spiral galaxy in the Virgo Cluster of galaxies. On good nights, with a large telescope, you may make out the wonderful spiral arms of this galaxy, because you see it face on.

Hydra

Hydra the Water Snake (see Figure 12-15) is the largest of all the constellations, snaking its way across 90 degrees of sky. This constellation has hardly any bright stars in it, though, and those it does have are spread out so much that it makes finding Hydra rather tricky.

The only really bright star in Hydra is *Alphard,* which lies in a relatively blank patch of sky between the equally dim constellations of Sextans and Monoceros. The head of Hydra is made up of a small loop of stars that sit just south of Cancer the Crab.

Look for the following in Hydra The numbers in the list correspond to the numbers in Figure 12-15:

1. **M48** is a good open cluster that you can find as the third corner of an equilateral triangle containing the head of Hydra and *Alphard,* much nearer the constellation Monoceros. You should be able to make it out with your naked eye, but binoculars will show it much more clearly.

2. **M68** is a globular cluster near the constellation Corvus. A line drawn from δ Corvi to β Corvi points to M68 in Hydra.

3. **M83** is a large, round spiral galaxy.

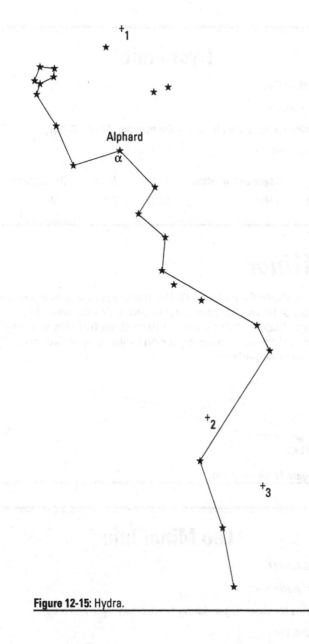

Figure 12-15: Hydra.

Hydra info

Abbreviation: Hya

Genitive: Hydrae

Best months for observing in during the evening: April and May

Bright Stars List

Name	Bayer Designation	Type	Mag	Distance (ly)
Alphard	α Hya	K3III	1.99	177

Leo Minor

Leo Minor (see Figure 12-16), the Lesser Lion, is a pale reflection of its brighter namesake, Leo. It lies between Leo and Ursa Major, so finding it is rather straightforward, even if the faint stars that comprise it don't join up to make an easily recognisable pattern.

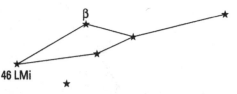

Figure 12-16: Leo Minor.

Leo Minor info

Abbreviation: LMi

Genitive: Leonis Minoris

Best months for observing in during the evening: April and May

Bright Stars List

Name	Bayer Designation	Type	Mag	Distance (ly)
Praecipua	46 LMi	K0III	3.79	98

Leo Minor has only a few hard-to-find galaxies. You'll need a medium or large telescope, a detailed star map and lots of patience to find this constellation.

Lupus

Lupus the Wolf (see Figure 12-17) is a southern hemisphere constellation lying north of α and β Centauri, between Centaurus and Scorpius. Lupus has many bright-ish stars, but the shape of a wolf is quite hard to trace.

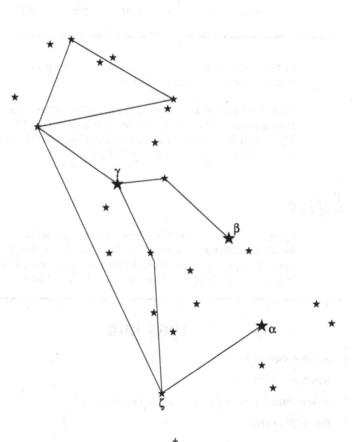

Figure 12-17: Lupus.

Lupus info

Abbreviation: Lup

Genitive: Lupi

Best months for observing in during the evening: May and June

Bright Stars List

Name	Bayer Designation	Type	Mag	Distance (ly)
	α Lup	B1III	2.30	548
	β Lup	B2III	2.68	523
	γ Lup	B2IV	2.80	567

Lupus appears in the Milky Way but has relatively few faint fuzzies to attract your attention.

Look for **NGC 5822,** an open cluster of stars that looks good through binoculars or a telescope. It's numbered 1 in Figure 12-17, and lies about three-quarters of the way along a line drawn between α Centauri and ξ Lupi.

Lynx

Lynx (see Figure 12-18) is a large and faint constellation located between Ursa Major and Auriga. Lynx doesn't have many bright stars in it, despite its size. You can find the brightest star in Lynx, α Lyncis, next to Leo Minor.

Lynx info

Abbreviation: Lyn

Genitive: Lyncis

Best months for observing in during the evening: March

Bright Stars List

Name	Bayer Designation	Type	Mag	Distance (ly)
	α Lyn	K7III	3.14	222

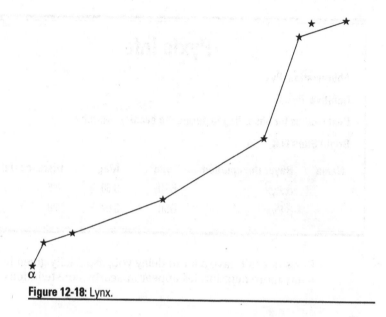

Figure 12-18: Lynx.

Lynx has very few faint fuzzies in its relatively blank bit of sky.

Pyxis

Pyxis the Compass (see Figure 12-19) is a tiny southern constellation made up of rather faint stars. It's sandwiched between Puppis and Vela, the Poop-deck and the Sail, just on the edge of the band of the Milky Way.

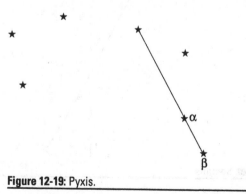

Figure 12-19: Pyxis.

Pyxis info

Abbreviation: Pyx

Genitive: Pyxidis

Best months for observing in during the evening: March

Bright Stars List

Name	Bayer Designation	Type	Mag	Distance (ly)
	α Pyx	B1III	3.68	845
	β Pyx	G5II	3.97	388

Pyxis doesn't have a lot to delay you, especially given how many more faint fuzzies appear in nearby constellations.

Sextans

Sextans the Sextant (see Figure 12-20) is another faint constellation sitting on the celestial equator, beside the bright star *Alphard,* α Hydrae.

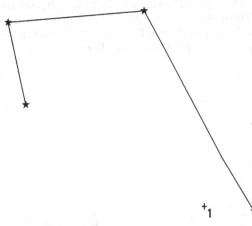

Figure 12-20: Sextans.

Sextans info

Abbreviation: Sex

Genitive: Sextantis

Best months for observing in during the evening: April

Sextans has no stars brighter than magnitude 4.

The **Spindle Galaxy, NGC 3115,** is a rather faint galaxy, which you can see through a medium telescope. It's numbered 1 in Figure 12-20. Draw a line from *Alphard* to γ Sextantis. Half that distance again is where you find the Spindle Galaxy.

Chapter 13

Stars of June, July and August

. . . .

In This Chapter

▶ Identifying the constellations visible in the northern hemisphere summer

▶ Identifying the constellations visible in the southern hemisphere winter

▶ Finding the interesting objects in these constellations

. . . .

Dark winter skies in the southern hemisphere reveal a host of stunning constellations such as Scorpius the Scorpion and Sagittarius the Archer. These two giants are easily visible from most of the US, southern Europe, the Middle East and south Asia, too.

Astronomers north of around 45 degrees north won't be able to see the entire constellations of Scorpius and Sagittarius, which will lie very low on the southern horizon. The bright stars of the Summer Triangle, an asterism made up of the brightest stars in the constellations of Cygnus, Lyra and Aquila, shine high in northern hemisphere skies.

Constellations of June, July and August

The constellations in this chapter are best seen in the months of June, July and August, and a map of these is shown in Figure 13-1.

REMEMBER

The chart above shows all the constellations in this chapter. Northern hemisphere stargazers may only be able to see the constellations at the top of this chart, while southern hemisphere stargazers may only be able to see those at the bottom. The horizon line marked in the figure shows what a stargazer at mid-northern latitudes will see.

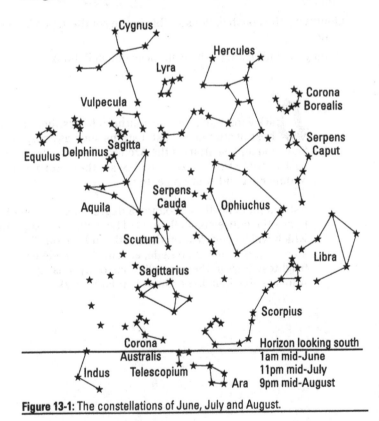

Figure 13-1: The constellations of June, July and August.

Remember that as the Earth spins, the position and orientation of the constellations changes. Figure 13-1 shows the relative positions of the constellations next to one another. Check monthly star maps for your location to find out where in the sky the constellations actually appear.

Cygnus

Cygnus the Swan (see Figure 13-2) is a majestic northern constellation, and you should really be able to see the figure of a swan drawn out in its stars. The tail of the swan is the bright star *Deneb,* and the head of the swan is *Albireo,* a great double star that you can make out through binoculars as one orange and one blue star.

Cygnus looks like it's flying along the Milky Way, which means that it's packed full of interesting faint fuzzies.

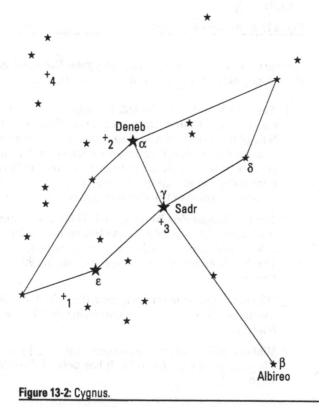

Figure 13-2: Cygnus.

Deneb, the brightest star in Cygnus, is one of the 20 brightest stars in the sky and part of the large asterism the Summer Triangle, shown in Figure 13-3. The other stars in this asterism are *Vega* in the constellation Lyra, and *Altair* in Aquila, both of which are brighter than *Deneb.*

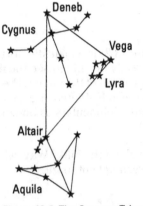

Figure 13-3: The Summer Triangle.

Here are a few things to look for in Cygnus The numbers in the list correspond to the numbers in Figure 13-2:

1. The **Veil Nebula, NGC 6992,** is a huge faint nebula made up of the gas from a supernova, a star that exploded tens of thousands of years ago. You may be able to find it using a telescope if you've got dark skies, but the Veil Nebula appears best in photographs. You can find it by continuing the line drawn from *Sadr,* γ Cygni, to ε Cygni by about a third of the distance again.

2. The **North America Nebula, NGC 7000,** is another large faint fuzzy, but this one is easy to find under dark skies using binoculars or a telescope with a wide field of view. The North America Nebula lies near the bright star *Deneb.*

3. **M29** is an open cluster lying near the bright star *Sadr,* γ Cygni. You'll see it through binoculars or a low-power telescope.

4. **M39** is a large and scattered open cluster, so you'll need a large field of view to find it. It lies behind the tail of the swan.

Cygnus info

Abbreviation: Cyg

Genitive: Cygni

Best months for observing in during the evening: August and September

Bright Stars List

Name	Bayer Designation	Type	Mag	Distance (ly)
Deneb	α Cyg	A2Ia	1.25	3228
Albireo	β Cyg	K3II	3.05	385
Sadr	γ Cyg	F8Ib	2.23	1523
	δ Cyg	B9III	2.86	171
Gienah	ε Cyg	K0III	2.48	72

Scorpius

Scorpius the Scorpion (see Figure 13-4) is one of the best zodiac constellations in the night sky. The bright red star *Antares* makes Scorpius easy to find, and the sweeping shape of the scorpion's tail – complete with stinger – makes one of the most distinct shapes in the sky. Scorpius is packed with bright stars, which makes finding the scorpion very easy.

Scorpius is a big constellation, but the ecliptic line passes through the narrow north part of Scorpius, near β Scorpii, which means that the planets, Moon and Sun don't spend much time in Scorpius. The Sun passes through Scorpius between 23 and 29 November.

Scorpius lies in part of the Milky Way near the galactic centre, close to the constellation Sagittarius, which means that you can see a lot of great faint fuzzies.

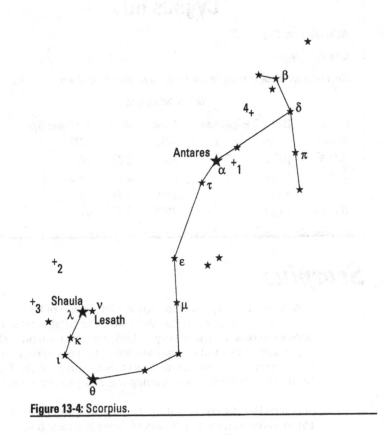

Figure 13-4: Scorpius.

Here are things to look for in Scorpius. The numbers in the list correspond to the numbers in Figure 13-4:

1. **M4** is a globular cluster lying in the same field of view as *Antares* when seen through binoculars, which makes it very easy to find.

2. The **Butterfly Cluster, M6,** is one of two open clusters just north of the bright star *Shaula,* λ Scorpii. It's more distant and therefore fainter than its neighbour, M7.

3. The **Ptolemy Cluster, M7,** looks like a fuzzy clump of stars to your naked eye, but if you look through binoculars, you'll be treated to a beautiful open cluster of stars.

4. **M80** is a globular cluster lying half way between *Antares* and *Graffias*. Through a medium-sized telescope, you'll see M80 as a fuzzy ball.

Scorpius info

Abbreviation: Sco

Genitive: Scorpii

Best months for observing in during the evening: July and August

Bright Stars List

Name	Bayer Designation	Type	Mag	Distance (ly)
Antares	α Sco	M1Ib	1.06	604
Graffias	β Sco	B05	2.56	530
	δ Sco	B0IV	2.29	401
	ε Sco	K2IIIB	2.29	65
	θ Sco	F1II	1.86	272
	ι Sco	F3Ia	2.99	1791
	κ Sco	B1III	2.39	464
Shaula	λ Sco	B1V	1.62	703
	μ Sco	B1IV	3.00	821
	π Sco	B1V	2.89	459
	τ Sco	B0V	2.82	430
Lesath	υ Sco	B2IV	2.70	518

Sagittarius

Sagittarius the Archer (see Figure 13-5) is another bright zodiac constellation, although it doesn't look especially like an archer. Maybe that's because the archer in question is a centaur (half man, half horse)!

Although you may not be able to recognise the archer, the brighter stars in Sagittarius are a bit easier to see. They make up an asterism known as the Teapot, with the Milky Way appearing to emerge from the spout like steam!

Sagittarius is a zodiac constellation, with the ecliptic running through it. The Sun is in Sagittarius between 18 December and 19 January.

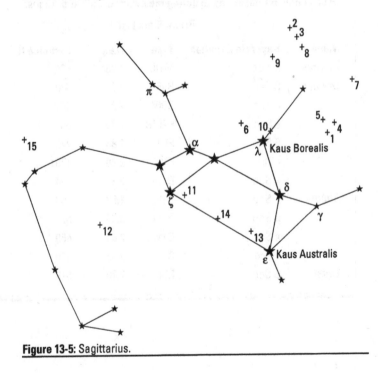

Figure 13-5: Sagittarius.

Sagittarius info

Abbreviation: Sgr

Genitive: Sagittarii

Best month for observing in during the evening: August

Bright Stars List

Name	Bayer Designation	Type	Mag	Distance (ly)
	γ Sgr	K0III	2.98	96
Kaus Media	δ Sgr	K3III	2.72	306
Kaus Australis	ε Sgr	B9III	1.79	145
Ascella	ζ Sgr	A3IV	2.60	89
Kaus Borealis	λ Sgr	K1III	2.82	77
	π Sgr	F2II	2.88	440
Nunki	σ Sgr	B2V	2.05	224

The Milky Way, the Earth's galaxy, is a band of faint light stretching across the sky, but it reaches its brightest when you look towards the centre of the galaxy. To see the Milky Way at its brightest, you should look towards Sagittarius. The fact that the centre of the Milky Way lies in this direction means that the sky in this constellation is one of the richest parts of the night sky anywhere.

Look for these things in Sagittarius. The numbers in the list correspond to the numbers in Figure 13-5:

1. The **Lagoon Nebula, M8,** is a naked-eye faint fuzzy, a cloud of interstellar gas (an emission nebula) lit up by the stars that are forming inside it, which looks good through binoculars or a telescope. It lies north of the spout of the Teapot.

2. The **Omega Nebula, M17,** is another gas cloud. Although you can view it using binoculars, to see it properly you need a telescope. The Omega Nebula is horseshoe-shaped, like the Greek capital letter Ω.

3. **M17** is an open cluster of stars, best seen using a telescope.

4. The **Trifid Nebula, M20,** is yet another emission nebula that you can see with a small telescope. It has a faint double star in the centre. If you observe the Trifid Nebula with a large telescope, you may be able to see the dust lanes that trisect this nebula.

5. **M21** is an open cluster that is visible through binoculars.

6. **M22** is a globular cluster, one of the brightest in the sky. You can just barely see it with the naked eye under perfect conditions, but you need a telescope to see the brighter stars within it.

7. **M23** is an open cluster that you can see best through a telescope, but you'll be able to find it using binoculars.

8. The **Sagittarius Star Cloud, M24,** is a large field of stars that looks great through binoculars.

9. **M25** is a faint open cluster.

10. **M28** is a globular cluster very near *Kaus Borealis*. If you look at it through a medium or large telescope, you should begin to resolve some stars in the fuzzy ball.

11. **M54** is a globular cluster near *Ascella*. It's hard to resolve into individual stars though, so it looks like a fuzzy ball even through a large telescope.

12. **M55** is a large globular cluster which you should be able to resolve into individual stars through a large telescope.

13. **M69** is another globular cluster, hard to resolve.

14. **M70** is a small and faint globular cluster.

15. **M75** is a very small, compact and bright globular cluster.

Aquila

Aquila the Eagle (see Figure 13-6) is an indistinct constellation with the bright star *Altair* in it. Like its neighbour Cygnus, Aquila sits in the Milky Way, and *Altair* is part of the Summer Triangle asterism.

Aquila info

Abbreviation: Aql

Genitive: Aquilae

Best months for observing during the evening: August and September

Bright Stars List

Name	Bayer Designation	Type	Mag	Distance (ly)
Altair	α Aql	A7IV	0.76	17
Alshain	β Aql	G8IV	3.71	45
Tarazed	γ Aql	K3II	2.72	460
	ζ Aql	A0V	2.99	83

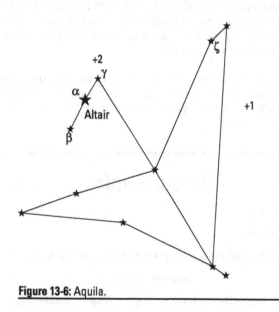

Figure 13-6: Aquila.

Altair, the brightest star in Aquila, is one of three stars, along with *Deneb* in Cygnus and *Vega* in Lyra, that form the asterism known as the Summer Triangle. (Refer to Figure 13-3 for a map.)

Here are a few things to look for in Aquila. The numbers in the list correspond to the numbers in Figure 13-6:

1. **NGC 6709** is a faint open cluster lying between Aquila and Hercules. If you look through a small telescope, you should be able to make out individual stars in the cluster.

2. The **E Nebula** is an unusual nebula; rather than shining brightly, the E Nebula is made up of dark gas obscuring starlight behind it. The sections of this nebula form the shape of a capital E. The nebula lies in the same binocular field of view as *Tarazed*.

Hercules

Hercules the Warrior (see Figure 13-7) is a big constellation, but because it doesn't have many bright stars in it, it can be sometimes difficult to locate. Hercules lies about one-third of the way along the line drawn between the bright stars *Vega* in Lyra and *Arcturus* in Boötes.

In Greek myth, Hercules was given 12 labours to perform, one of which was to slay a dragon, represented by the constellation Draco, which sits north of Hercules, nearer the North Star.

Hercules info

Abbreviation: Her

Genetive: Herculis

Best months for observing in during the evening: July and August

Bright Stars List

Name	Bayer Designation	Type	Mag	Distance (ly)
Rasalgethi	α Her	M5II	3.31	382
Kornophoros	β Her	G8III	2.78	148
	γ Her	A9III	3.74	195
	ζ Her	F9IV	2.81	35

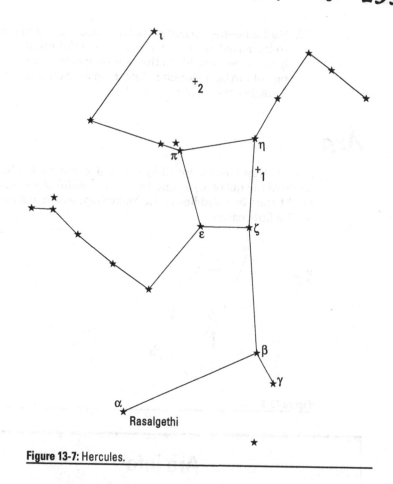

Figure 13-7: Hercules.

Hercules' body is marked by four stars – ε, ζ, η and π Herculis – in a quadrilateral asterism known as the *Keystone*.

Here are a few things to look for in Hercules The numbers in the list correspond to the numbers in Figure 13-7:

1. **Great Hercules Cluster, M13,** is the brightest globular cluster in the northern hemisphere. You can find it two-thirds of the way along a line drawn between ζ and η Herculis, which are two of the stars of the Keystone. The Great Hercules Cluster is visible to the naked eye as a fuzzy star, but a telescope enables you to see the individual stars within the cluster.

2. **M92** is another bright globular cluster, just about visible to the naked eye under very good conditions. It's often ignored because it's in the same constellation as M13, one of the best globular clusters of all. You can find it on a line between η and ι Herculis.

Ara

Ara the Altar (see Figure 13-8) is a small constellation lying south of the tail of Scorpius. Ara has a handful of moderately bright stars in it and lies in the Milky Way, which makes finding it a little easier.

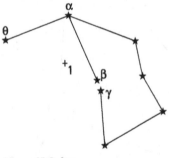

Figure 13-8: Ara.

Ara info

Abbreviation: Ara

Genitive: Arae

Best months for observing in during the evening: July and August

Bright Stars List

Name	Bayer Designation	Type	Mag	Distance (ly)
	α Ara	B2V	2.84	242
	β Ara	K3Ib	2.84	603
	γ Ara	B1Ib	3.31	1136

Look for **NGC 6397,** a globular cluster in which the stars are a bit more spread out than normal. It's numbered 1 in Figure 13-8. NGC 6397 is one of the closest globular clusters to Earth, lying only 10,000 light years away, and is just visible to the naked eye as a fuzzy star. NGC 6397 lies a little less than half way between β Arae and θ Arae.

Corona Australis

Corona Australis, the Southern Crown, (see Figure 13-9) is a smaller version of its northern namesake, Corona Borealis, with an attractive curve of stars representing a wreath. Corona Australis sits beneath the Teapot asterism in Sagittarius and beside the tail of Scorpius.

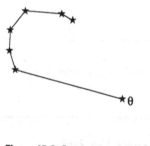

Figure 13-9: Corona Australis.

Look for **NGC 6541,** a globular cluster and the only decent faint fuzzy in Corona Australis. It's numbered 1 in Figure 13-9. You can see it with a small telescope about half way between θ Coronae Australis and θ Scorpii.

Corona Australis info

Abbreviation: CrA

Genitive: Coronae Australis

Best month for observing in during the evening: August

Corona Australis has no stars brighter than a magnitude 4.

Corona Borealis

Corona Borealis, the Northern Crown (see Figure 13-10), is a distinct semi-circle of stars between Hercules and Boötes. The bright-ish star *Alpheca* stands out in the base of the crown.

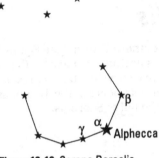

Figure 13-10: Corona Borealis.

Like nearby Boötes, this area of sky is pretty much empty of any faint fuzzies.

Corona Borealis info

Abbreviation: CrB

Genitive: Coronae Borealis

Best month for observing in during the evening: July

Bright Stars List

Name	Bayer Designation	Type	Mag	Distance (ly)
Alphecca	α CrB	A0V	2.22	75
Nusakan	β CrB	F0	3.66	14
	γ CrB	A1V	3.81	145

Delphinus

Delphinus the Dolphin (see Figure 13-11) is a tiny constellation, but very distinct despite its size, looking like a tiny dolphin leaping through space. You can find Delphinus between *Altair* in Aquila and the Great Square of Pegasus (see Chapter 14).

You won't find any bright faint fuzzies in this small constellation.

Figure 13-11: Delphinus.

Delphinus info

Abbreviation: Del

Genitive: Delphini

Best months for observing in during the evening: August and September

Bright Stars List

Name	Bayer Designation	Type	Mag	Distance (ly)
Sualocin	α Del	B9V	3.77	241
Rotanev	β Del	F5IV	3.64	97

Equuleus

Equuleus the Foal (see Figure 13-12) is tiny, the second small-est of all the constellations. With no very bright stars in it, Equuleus is not that easy to identify. Equuleus lies between Delphinus and Pegasus (see Chapter 14).

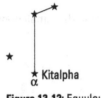

Kitalpha
α

Figure 13-12: Equuleus.

You won't find much to detain you in Equuleus.

Equuleus info

Abbreviation: Equ

Genitive: Equulei

Best months for observing in during the evening: August and September

Bright Stars List

Name	Bayer Designation	Type		Distance (ly)
Kitalpha	α Equ	GOIII	3.92	186

Indus

Indus the Indian (see Figure 13-13) is another indistinct constellation of the southern hemisphere. You can find it on a line between the brighter stars *Peacock,* α α Pavonis, and *Al Na'ir,* α Gruis.

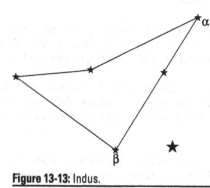

Figure 13-13: Indus.

Indus doesn't contain any bright faint fuzzies.

Indus info

Abbreviation: Ind

Genitive: Indi

Best month for observing in during the evening: August

Bright Stars List

Name	Bayer Designation	Type	Mag	Distance (ly)
	α Ind	K0III	3.11	101
	β Ind	K0III	3.67	603

Libra

Libra the Scales (see Figure 13-14) is one of the zodiac constellations. It sits between Scorpius and Libra, along a line between *Antares,* α Scorpii, and *Spica,* α Virginis. The stars of Libra aren't especially bright, so seeing the shape of the scales is much harder.

In fact, the stars of Libra were seen by ancient Greek astronomers as forming the pincers of Scorpius, but later Roman astronomers decided they formed the scales held by Virgo.

Because Libra was originally part of Scorpius, the brighter stars of Libra have rather exotic names: *Zubenelgenubi* and *Zubeneschamali,* the southern and the northern claw respectively.

Libra is on the ecliptic, and so the planets, Moon and Sun pass through here. The Sun is in Libra between 31 October and 23 November.

You won't find any bright faint fuzzies in Libra.

Libra info

Abbreviation: Lib

Genitive: Librae

Best months for observing in during the evening: June and July

Bright Stars List

Name	Bayer Designation	Type	Mag	Distance (ly)
Zubenelgenubi	α Lib	A3IV	2.75	77
Zubeneschamali	β Lib	B8V	2.61	160
	γ Lib	K0III	3.91	152

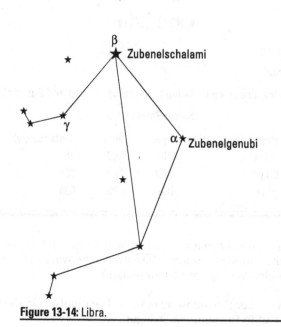

Figure 13-14: Libra.

Lyra

Lyra the Lyre (see Figure 13-15) is a tiny constellation made immediately recognisable by its brightest star, *Vega*, the fifth brightest star in the sky. Vega is brilliant white and is one of the closest stars to Earth, only 25 light years away.

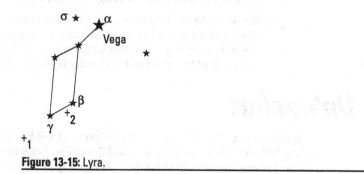

Figure 13-15: Lyra.

Lyra info

Abbreviation: Lyr

Genitive: Lyrae

Best months for observing in during the evening: August and September

Bright Stars List

Name	Bayer Designation	Type	Mag	Distance (ly)
Vega	α Lyr	A0V	0.03	25
Sheliak	β Lyr	A8V	3.52	881
Sulafat	γ Lyr	B9III	3.25	634

Vega is one of three stars, along with *Deneb* in Cygnus and *Altair* in Aquila, that form the asterism known as the Summer Triangle. (See Figure 13-3 for a map.)

Look out for the following in Lyra. The numbers in the list correspond to the numbers in Figure 13-15:

1. **M56,** a globular cluster that you can resolve into individual stars with a large telescope. Find it by drawing a line from *Sheliak* through *Sulafat* and continuing a little more than the same distance again.

2. The Ring **Nebula, M57,** is an oval cloud of gas, a so-called planetary nebula. The gas in this cloud was puffed off by a dying star in its final death throes. You can see the oval shape through a small telescope.

3. The star ε Lyrae is known as the Double Double star. When you look at ε Lyrae using binoculars you see a double star, but if you use a telescope you can see that each of the stars in the double is itself a double!

Ophiuchus

Ophiuchus the Serpent-bearer (see Figure 13-16) is a giant constellation between Hercules and Scorpius. Because of its size, joining the dots in Ophiuchus is often quite difficult, but the constellation is meant to represent a man carrying a snake!

Ophiuchus actually straddles the ecliptic, making it a zodiac constellation, albeit not one of the famous 12. The Sun passes through Ophiuchus between 30 November and 17 December.

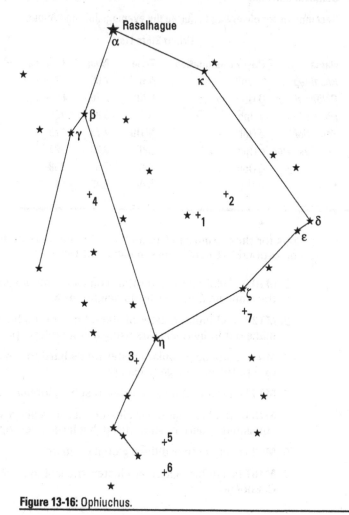

Figure 13-16: Ophiuchus.

Ophiuchus info

Abbreviation: Oph

Genitive: Ophiuchi

Best months for observing in during the evening: July and August

Bright Stars List

Name	Bayer Designation	Type	Mag	Distance (ly)
Rasalhague	α Oph	A5III	2.08	47
Celbalrai	β Oph	K2III	2.76	82
Marfik	γ Oph	A2V	3.82	166
Yed Prior	δ Oph	M1III	2.73	170
Yed Posterior	ε Oph	G8III	3.23	107
	ζ Oph	O9V	2.54	458
Sabik	η Oph	A2V	2.43	84

Look for these things in Ophiuchus. The numbers in the list correspond to the numbers in Figure 13-16:

1. **M10** is a globular cluster that you can sometimes get in the same field of view as its neighbour M12.

2. **M12** is a globular cluster of stars. You should be able to make out individual stars using a large telescope.

3. **M9** is a diffuse globular cluster that's hard to resolve except through large telescope.

4. **M14** is another diffuse, hard-to-resolve globular cluster.

5. **M19** is an oblate globular cluster – it looks like a slightly squashed sphere of stars through a large telescope.

6. **M62** is yet another diffuse globular cluster.

7. **M107** is a diffuse globular cluster, the last in the Messier Catalogue.

Scutum

Scutum the Shield (see Figure 13-17) is a small indistinct constellation that lies in the Milky Way, which makes it slightly easier to find. Scutum lies south of Aquila the Eagle and has several faint fuzzies; one in particular looks especially good through a telescope.

Look for these in Scutum. The numbers in the list correspond to the numbers in Figure 13-17:

1. The **Wild Duck Cluster, M11,** is a great open cluster. Through a small telescope, the V-shape of stars looks like ducks in flight. You can find the Wild Duck Cluster near the middle of the thin triangle made up from β, δ and η Scuti.

2. **M26** is a small, faint open cluster just next to δ Scuti.

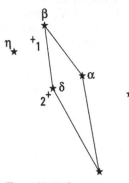

Figure 13-17: Scutum.

Scutum info

Abbreviation: Sct

Genitive: Scuti

Best month for observing in during the evening: August

Bright Stars List

Name	Bayer Designation	Type	Mag	Distance (ly)
	α Sct	K2III	3.85	174

Serpens

Serpens the Serpent is a unique constellation. It's made up of two separate parts, one on either side of Ophiuchus the Serpent-bearer. The two parts are known as Serpens Caput and Serpens Cauda (the tail and head, respectively).

Serpens Caput

Serpens Caput (see Figure 13-18) has the brighter stars in it and lies between Boötes and Ophiuchus.

Look for **M5,** one of the better globular clusters in the sky (numbered 1 in Figure 13-18). To find M5, pretend that the faint stars ε and μ Serpentis are a mirror; M5 is at the mirror-image position from the faint σ Serpentis.

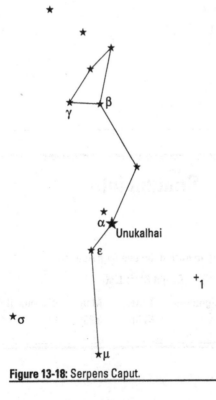

Figure 13-18: Serpens Caput.

Serpens info

Abbreviation: Ser

Genitive: Serpentis

Best months for observing in during the evening: July and August

Bright Stars List

Name	Bayer Designation	Type	Mag	Distance (ly)
Unukalhai	α Ser	K2III	2.63	73
	β Ser	A3V	3.65	153
	γ Ser	F6V	3.85	36

Serpens Cauda

Serpens Cauda (see Figure 13-19) is much dimmer than Serpens Caput and harder to find. Serpens Cauda lies between Ophiuchus and Scutum, which are themselves quite hard to locate.

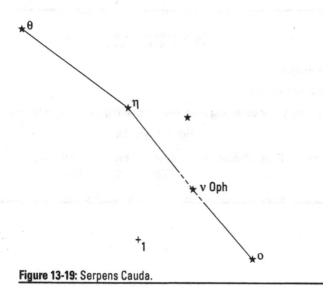

Figure 13-19: Serpens Cauda.

Look out for **M16,** a fuzzy open cluster lying in front of the Eagle Nebula. It's numbered 1 in Figure 13-19. You can see M16 through binoculars, but you need a large telescope to make out the gas in the Eagle Nebula.

Sagitta

Sagitta the Arrow (see Figure 13-20) is a tiny constellation lying between the constellations Aquila and Cygnus. The stars in it are all quite dim, but the four brightest do make up a prominent, if small, arrow shape.

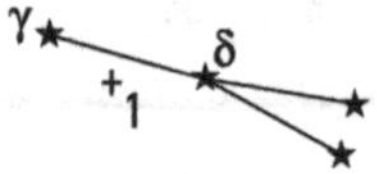

Figure 13-20: Sagitta.

Look for **M71,** a globular cluster lying between γ Sagittae and δ Sagittae, the two brightest stars in this little constellation. It's numbered 1 in Figure 13-20.

Sagitta info

Abbreviation: Sge

Genitive: Sagittae

Best months for observing in during the evening: August and September

Bright Stars List

Name	Bayer Designation	Type	Mag	Distance (ly)
	γ Sge	K5III	3.51	274

Telescopium

Telescopium the Telescope (see Figure 13-21) is another dim and indistinct constellation of the southern hemisphere. Telescopium lies south of Scorpius and Sagittarius. Three stars appear in a right-angled triangle, and the rest of the constellation straggles back towards the star *Peacock,* α Pavonis

You won't see great deal to interest you in Telescopium; it doesn't live up to its name!

Figure 13-21: Telescopium.

Telescopium info

Abbreviation: Tel

Genitive: Telescopii

Best months for observing in during the evening: August

Bright Stars List

Name	Bayer Designation	Type	Mag	Distance (ly)
	α Tel	B3IV	3.49	249

Vulpecula

Vulpecula the Fox (see Figure 13-22) lies south of Cygnus, near the bright star *Albireo,* α Cygni, and is generally faint and indistinct, with no bright stars in it. Vulpecula lies in the Milky Way and so is a rich field for finding faint fuzzies.

Look for these in Vulpecula. The numbers in the list correspond to the numbers in Figure 13-22:

1. The **Dumbbell Nebula, M27,** is a twin-lobed planetary nebula, which is a great telescope object. It's also visible through binoculars. The Dumbbell Nebula makes an equilateral triangle with the very faint stars 12 and 13 Vulpeculae.

2. **Brocchi's Cluster** is an unusual faint fuzzy in that it's made up of ten stars – six in a line and four in a hook hanging from that line, which earned the cluster the name the Coathanger. You can easily see Brocchi's Cluster through binoculars; find it between 1 and 9 Vulpeculae.

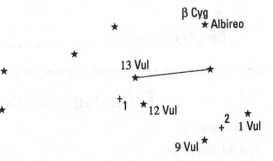

Figure 13-22: Vulpecula.

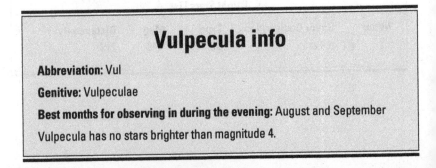

Vulpecula info

Abbreviation: Vul

Genitive: Vulpeculae

Best months for observing in during the evening: August and September

Vulpecula has no stars brighter than magnitude 4.

Chapter 14

Stars of September, October and November

● ●

In This Chapter

▷ Identifying the constellations visible in the northern hemisphere autumn

▷ Identifying the constellations visible in the southern hemisphere spring

▷ Finding the interesting objects in these constellations

● ●

he stars of Andromeda, Perseus and Pegasus dominate the skies during September, October and November for most stargazers, except those south of around 30 degrees south. Andromeda features one of the best faint fuzzies in the sky, the Andromeda Galaxy.

Constellations of September, October and November

The constellations in this chapter are best seen in the months of September, October and November, and a map of these is shown in Figure 14-1.

Figure 14-1 shows all the constellations in this chapter. Northern hemisphere stargazers may only be able to see the constellations at the top of this chart, while southern hemisphere stargazers may only be able to see those at the bottom. The horizon line marked in the figure shows what a stargazer at mid-northern latitudes will see.

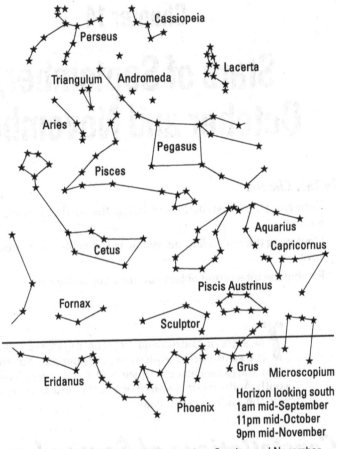

Figure 14-1: The constellations of September, October and November.

Remember that as the Earth spins, the position and orientation of the constellations changes. Figure 14-1 shows the relative positions of the constellations next to one another. Check monthly star maps for your location to find out where in the sky the constellations actually appear.

Andromeda

Andromeda (see Figure 14-2) represents the princess daughter of Cassiopeia and Cepheus, who was left as a sacrifice for the sea monster Cetus, but who was ultimately rescued by the hero Perseus.

Andromeda lies between Pegasus and Perseus in the sky and has three bright stars in it, in a shallow curve.

The giant Andromeda Galaxy is huge, larger than the Milky Way, containing close to a trillion stars (1,000,000,000,000!), but from a position 2.5 million light years distant, it looks like a small smudge in the sky. Andromeda is hurtling towards the Milky Way, and in a few billion years these two galaxies will collide, perhaps merging to form a giant galaxy.

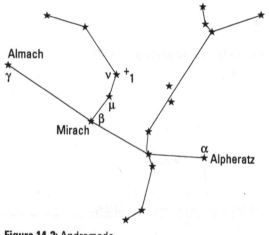

Figure 14-2: Andromeda.

Andromeda info

Abbreviation: And

Genitive: Andromedae

Best months for observing in during the evening: October and November

Bright Stars List

Name	Bayer Designation	Type	Mag	Distance (ly)
Alpheratz	α And	B9	2.07	97
Mirach	β And	MOIII	2.07	199
Almach	γ And	B8V	2.10	355

You can find the Andromeda Galaxy (see Figure 14-3) by using either Cassiopeia or Pegasus to locate the star *Mirach,* β Andromedae, which is the middle of the three bright stars in the curve of Andromeda. Then you should step up via μ Andromedae to a faint fuzzy patch in the sky.

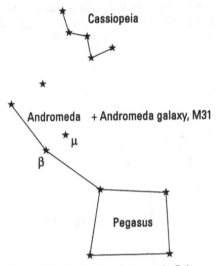

Figure 14-3: Finding the Andromeda Galaxy.

Look for the **Andromeda Galaxy, M31,** one of the best faint fuzzies in the sky: it's numbered 1 in Figure 14-2. You can see the Andromeda Galaxy, M31, as an elongated smudge with your naked eye under dark skies, and binoculars will enlarge this smudge. With a medium telescope, you can see more detail, including the companion galaxies that lie near the Andromeda Galaxy – M32 and M110.

Pegasus

Pegasus the Winged Horse (see Figure 14-4) was believed to have sprung from the blood of the Gorgon Medusa, which spilled into the sea after Perseus cut off Medusa's head.

The constellation of Pegasus is made up of a prominent square of stars, one of which is technically in the constellation of Andromeda (see preceding section).

Pegasus info

Abbreviation: Peg

Genitive: Pegasi

Best month for observing in during the evening: October and November

Bright Stars List

Name	Bayer Designation	Type	Mag	Distance (ly)
Markab	α Peg	B9III	2.49	140
Scheat	β Peg	M2II	2.44	199
Algenib	γ Peg	B2IV	2.83	333
Enif	ε Peg	K2Ib	2.38	672
Matar	η Peg	G2II	2.93	215

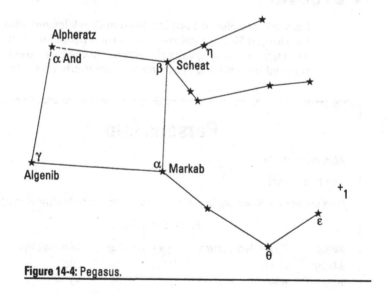

Figure 14-4: Pegasus.

Pegasus lies at one end of the constellation Andromeda, between that constellation and faint Aquarius.

The asterism known as the Great Square of Pegasus is made up of the stars *Markab, Scheat* and *Algenib* in Pegasus, plus the star *Alpheratz* in Andromeda. If you're stargazing in a town or city, The Great Square of Pegasus may look like an empty square, with nothing much inside it. Under darker skies, though, you can make out a handful of stars of magnitudes between 4.5 and 5.5.

The number of stars you can count inside the Great Square of Pegasus tells you how dark your skies are.

Look for **M15,** a good globular cluster that you can find by continuing the line between θ and ε Pegasi by about half that distance again. It's numbered 1 in Figure 14-4. You can see M15 using binoculars, and you can make out stars with a medium telescope.

Perseus

Perseus (see Figure 14-5) is a hero in Greek legend who slew the Gorgon Medusa (whose face turned people to stone, and from whose blood sprang the winged horse Pegasus) and who rescued the princess Andromeda from the sea monster Cetus.

Perseus info

Abbreviation: Per

Genitive: Persei

Best months for observing in during the evening: November and December

Bright Stars List

Name	Bayer Designation	Type	Mag	Distance (ly)
Mirfak	α Per	F5Ib	1.79	592
Algol	β Per	B8V	2.09	93
	γ Per	G8III	2.91	256
	ε Per	B0V	2.90	538
	ζ Per	B1Ib	2.84	982

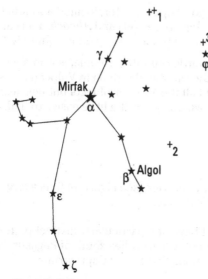

Figure 14-5: Perseus.

You can find Perseus between Auriga and Andromeda, in the Milky Way next to Cassiopeia.

Perseus isn't quite as distinct as the nearby Cassiopeia, as the stars in it aren't quite as bright, but it has a rather attractive sweeping curve of stars: α, γ, δ, η, λ and μ Persei.

Algol, β Persei, is known as the Demon Star (El Ghoul) and is an *eclipsing binary* – two stars that orbit their common centre of gravity and make one orbit every three days. As the fainter star passes in front of the brighter one, it blocks out a bit of the brighter star's light, which makes the brightness you see on Earth dim a little. This means that *Algol* varies in brightness. Usually, *Algol's* magnitude is 2.1, but every 2.87 days, it dims to magnitude 3.4, going from among the 60 brightest stars to only one of the top 300 stars.

Look for these things in Perseus. The numbers in the list correspond to the numbers in Figure 14-5:

1. The **Sword Handle, NGC 869** and **884,** is a great pair of open clusters, visible to the naked eye as side-by-side smudges. You can find it in the Milky Way half way between the bright stars of γ Persei and *Ruchbah* (in Cassiopeia).

2. **M34** is a good open cluster, located about half way between *Algol* (β Persei) and *Almach*. You can see M34 easily through binoculars, even in light-polluted skies.

3. **M76,** the Little Dumbbell Nebula, is a smaller version of M27, the Dumbbell Nebula in Vulpecula. It's the faintest of all the Messier objects but lies near the star φ Persei, which makes it a little easier to find.

Aquarius

Aquarius the Water-carrier (see Figure 14-6) is a large constellation lying on the zodiac.

Aquarius doesn't have any particularly distinct patterns to make it easy to find. Aquarius lies south of Pegasus, between the dim constellations Pisces and Capricornus.

Aquarius is one of the zodiac constellations, so the planets, Moon and Sun move through it regularly. The Sun is in Aquarius between 16 February and 11 March every year.

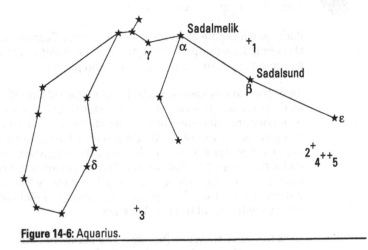

Figure 14-6: Aquarius.

Aquarius info

Abbreviation: Aqr

Genitive: Aquarii

Best month for observing in during the evening: October

Bright Stars List

Name	Bayer Designation	Type	Mag	Distance (ly)
Sadalmelik	α Aqr	G2Ib	2.95	758
Sadalsuud	β Aqr	G0Ib	2.90	612
Sadachbia	γ Aqr	A0V	3.86	158

Look for these things in Aquarius. The numbers in the list correspond to the numbers in Figure 14-6:

1. **M2** is a globular cluster just north of *Sadalsuud*. You see M2 looking like a fuzzy star in binoculars, but you need a medium telescope to resolve any stars.

2. The **Saturn Nebula, NGC 7009,** is a planetary nebula which, when seen through a large telescope, looks a bit like the planet Saturn, with lobes on either side, hence the name. You can find the Saturn Nebula between one-third and a quarter of the way along a line drawn between ε Aquarii and *Deneb Algedi,* but it's faint so will require some hunting!

3. The **Helix Nebula, NGC 7293,** is the largest of all planetary nebulae. It's a shell of gas puffed off by a dying star; through binoculars or a small telescope, the Helix Nebula appears like a faint grey ring. The Helix Nebula is located near the middle of a triangle of stars, δ Aquarii, *Deneb Algedi*, and *Fomalhaut,* a little closer to δ Aquarii than the others.

4. **M73** isn't actually a faint fuzzy at all. In fact it's just a collection of four faint stars next to one another that look a bit fuzzy through a small telescope, which is probably why Messier included them in his catalogue.

5. **M72** is a faint and unimpressive globular cluster.

Aries

Aries the Ram (see Figure 14-7) is a small zodiac constellation, with three bright-ish stars – α, β and γ Arietis – forming a slightly curved line. Aries lies between Taurus and Triangulum.

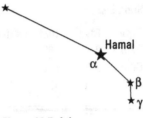

Figure 14-7: Aries.

Aries is on the ecliptic, so the Sun passes through this constellation between 18 April and 14 May.

You won't see any bright faint fuzzies in Aries.

Aries info

Abbreviation: Ari

Genitive: Arietis

Best months for observing in during the evening: November and December

Bright Stars List

Name	Bayer Designation	Type	Mag	Distance (ly)
Hamal	α Ari	K2III	2.01	66
Sheratan	β Ari	A5V	2.64	60
Mesarthim	γ Ari	B9V	3.88	204

Capricornus

Capricornus the Sea Goat (see Figure 14-8) is meant to represent a half-goat, half-fish sea creature.

This faint zodiac constellation isn't that easy to find; you
can track it down using the Summer Triangle, whose tip, the
bright star *Altair*, α Aquilae, points to *Dabih*, β Capricorni.

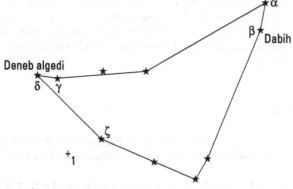

Figure 14-8: Capricornus.

Capricornus info

Abbreviation: Cap

Genitive: Capricorni

Best months for observing in during the evening: September and October

Bright Stars List

Name	Bayer Designation	Type	Mag	Distance (ly)
Algedi	α Cap	G8III	3.58	109
Dabih	β Cap	A5	3.05	344
Nashira	γ Cap	A7III	3.69	139
Deneb Algedi	δ Cap	A5	2.85	39

Capricornus is one of the zodiac constellations, which means that the planets, Moon and Sun pass through. The Sun lies in Capricornus between 19 January and 16 February.

Look for **M30,** an open cluster of stars that you can find through binoculars and which looks better through a telescope. It makes one point of a narrow wedge triangle where the tip is *Deneb Algedi, δ* Capricorni, and the third point is ζ Capricorni. It's numbered 1 in Figure 14-8.

Algedi Secunda and *Dabih* are nice binary stars, with *Algedi Secunda* being just a line-of-sight effect (the two stars lie in the same direction, but one is 500 light years farther away), while *Dabih* is a physical double star, with both stars orbiting each other.

Cetus

Cetus is the sea monster (see Figure 14-9) from whose maw the hero Perseus rescued the princess Andromeda.

Cetus lies south of Aries and is made up of a ring of stars for the tail (including *Menkar*) joined to the body (including *Diphda,* the brightest star in the constellation).

Cetus info

Abbreviation: Cet

Genitive: Ceti

Best months for observing in during the evening: November and December

Bright Stars List

Name	Bayer Designation	Type	Mag	Distance (ly)
Menkar	α Cet	M2III	2.54	220
Diphda	β Cet	K0III	2.04	96
	γ Cet	A3V	3.47	82
Mira	o Cet	M5e	3.04v	418

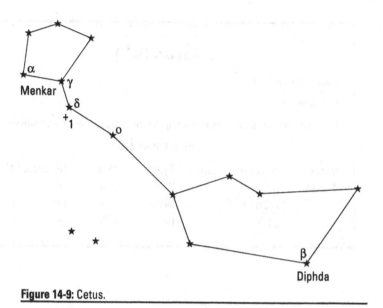

Figure 14-9: Cetus.

Look for **M77**, a face-on spiral galaxy located very near δ Ceti, which makes it easy to find. It's numbered 1 in Figure 14-9. M77 is best seen through a telescope and has a bright centre, which makes it look like just a fuzzy star.

Another interesting object in Cetus is the star *Mira,* o Ceti, which is a variable star whose brightness varies over the course of 332 days, between magnitude 2 and magnitude 10, which means that sometimes it's very bright indeed, and at other times it's not visible at all!

Grus

Grus the Crane (see Figure 14-10) is a southern constellation with a couple of bright stars in it, the brightest of which, *Al Na'ir,* α Gruis, is the thirtieth brightest star in the sky. The other stars in Grus are much fainter, which means tracing out the shape of this flying bird isn't nearly as easy as for the similarly shaped – and much larger – Cygnus in the northern skies.

Grus info

Abbreviation: Gru

Genitive: Gruis

Best months for observing in during the evening: October and November

Bright Stars List

Name	Bayer Designation	Type	Mag	Distance (ly)
Al Na'ir	α Gru	B7IV	1.73	101
	β Gru	M5III	2.07	170
	γ Gru	B8III	3.00	203

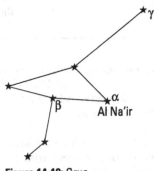

Figure 14-10: Grus.

You can find Grus south of Piscis Austrinus and the bright star *Fomalhaut*.

Grus doesn't have any bright faint fuzzies to look for.

Lacerta

Lacerta the Lizard (see Figure 14-11) is a rather small, undistinguished constellation between Andromeda and Cygnus in the northern skies.

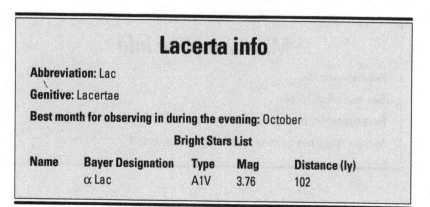

Lacerta info

Abbreviation: Lac

Genitive: Lacertae

Best month for observing in during the evening: October

Bright Stars List

Name	Bayer Designation	Type	Mag	Distance (ly)
	α Lac	A1V	3.76	102

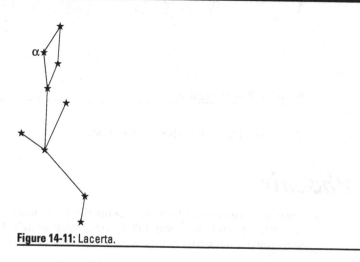

Figure 14-11: Lacerta.

Lacerta contains nothing that will keep you very long.

Microscopium

Microscopium the Microscope (see Figure 14-12) is a faint, indistinct constellation of the southern skies. Microscopium lies between the Teapot asterism in Sagittarius and the constellation Grus.

Microscopium info

Abbreviation: Mic

Genitive: Microscopii

Best months for observing in during the evening: September and October

Microscopium has no stars brighter than magnitude 4.

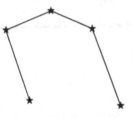

Figure 14-12: Microscopium.

You won't see a lot in this microscope.

Phoenix

Phoenix (see Figure 14-13) represents the fiery bird of legend which is reborn in fire every 1,000 years. Unfortunately, the constellation Phoenix isn't nearly as exciting as the name suggests!

Phoenix is rather hard to find, with only a few middle-brightness stars in it. Phoenix lies next to the bright star *Achernar,* α Eridani, one of the ten brightest stars in the sky, which helps when trying to find Phoenix.

Phoenix info

Abbreviation: Phe

Genitive: Phoenicis

Best month for observing in during the evening: November

Bright Stars List

Name	Bayer Designation	Type	Mag	Distance (ly)
Ankaa	α Phe	KOIII	2.40	77
	β Phe	G8III	3.32	198
	γ Phe	K5II	3.41	234

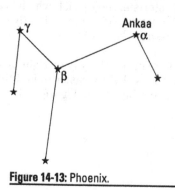

Figure 14-13: Phoenix.

Phoenix has no good faint fuzzies to keep you interested.

Pisces

Pisces the Fishes (see Figure 14-14) is one of the zodiac constellations, but it's notoriously hard to make out. Pisces covers a large area of sky between Aries and Aquarius, wrapping around the south-east corner of Pegasus.

Pisces info

Abbreviation: Psc

Genitive: Piscium

Best months for observing in during the evening: October and November

Bright Stars List

Name	Bayer Designation	Type	Mag	Distance (ly)
Alrescha	γ Psc	G7III	3.70	131
	η Psc	G8III	3.62	294

Pisces contains a faint asterism, the Circlet, which is a small ring of faint stars – γ, θ, ι, κ and λ Piscium – directly south of the Great Square of Pegasus.

Look for **M74,** a very faint spiral galaxy lying right beside η Piscium, which makes Pisces slightly easier to find than it would be otherwise. It's numbered 1 in Figure 14-14.

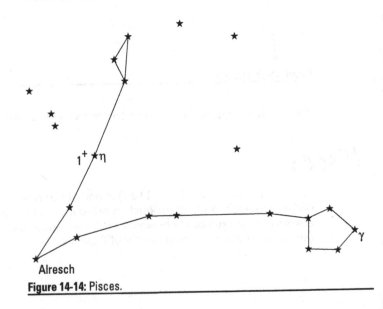

Figure 14-14: Pisces.

Piscis Austrinus

Piscis Austrinus the Southern Fish (see Figure 14-15) is the counterpart to Pisces the Fishes, who were said to be its offspring. Piscis Austrinus has only one bright star in it – *Fomalhaut,* α Piscis Austrini – which puts it in the top 20 brightest stars in the sky.

Piscis Austrinus lies between Aquarius (who is said to be pouring water into the fish's mouth) and Grus.

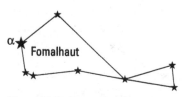

Figure 14-15: Pisces Austrinus.

You won't find any significant faint fuzzies in Piscis Austrinus.

Piscis Austrinus info

Abbreviation: PsA

Genitive: Piscis Austrini

Best month for observing in during the evening: November

Bright Stars List

Name	Bayer Designation	Type	Mag	Distance (ly)
Fomalhaut	α PsA	A3V	1.17	25

Sculptor

Sculptor (see Figure 14-16) is a dim constellation and not easy to identify. Sculptor lies between Piscis Austrinus and Fornax, themselves quite hard to locate.

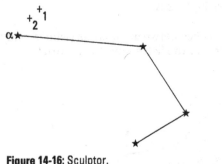

Figure 14-16: Sculptor.

Look for the following in Sculptor. The numbers in the list correspond to the numbers in Figure 14-16:

1. **NGC 253** lies next to NGC 253. It's tilted nearly edge-on to Earth, and so through a telescope you see a cigar-shaped smudge.

2. **NGC 288** is a globular cluster that you can see through a telescope. It lies next to the faint star α Sculptoris and in the field of view of NGC 253.

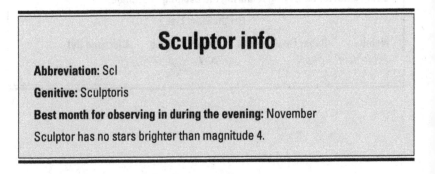

Sculptor info

Abbreviation: Scl

Genitive: Sculptoris

Best month for observing in during the evening: November

Sculptor has no stars brighter than magnitude 4.

Triangulum

The name Triangulum the Triangle (see Figure 14-17) is nothing if not descriptive. Triangulum's three faint stars lie between Andromeda and Aries, forming a small faint triangle in the sky.

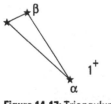

Figure 14-17: Triangulum.

Look for the **Triangulum Galaxy, M33,** a naked-eye galaxy – just. It's visible only if you have very dark skies. In fact, the Triangulum Galaxy is one of the indicators of a dark sky used in the Bortle Scale. It's only easily visible in Bortle Scale class 1 and 2 skies, but you may catch a glimpse of it with averted vision in rural Bortle class 3 skies. (See Chapter 2 for more on the Bortle Scale.)

Triangulum info

Abbreviation: Tri

Genitive: Trianguli

Best month for observing in during the evening: December

Bright Stars List

Name	Bayer Designation	Type	Mag	Distance (ly)
Mothallah	α Tri	F6IV	3.42	64
	β Tri	A5III	3.00	124

Chapter 15

Southern Polar Constellations

. . . .

In This Chapter

▶ Identifying the southern polar constellations

▶ Finding the interesting objects in these constellations

. . . .

Southern hemisphere stargazers are treated to many striking constellations that are visible any night of the year – the southern polar constellations.

These constellations, when observed from certain southern latitudes, are what's called *circumpolar*, meaning they never rise or set. That's very handy, since it means they are above the horizon all year, and are visible all night, so you'll soon get used to finding them and using them as signposts to other constellations.

Southern Polar Constellations

The stars around the southern pole form some striking constellations, such as the Southern Cross, and there are a wealth of great objects to look for, like the Large and Small Magellanic Clouds. Figure 15-1 is a map of these constellations.

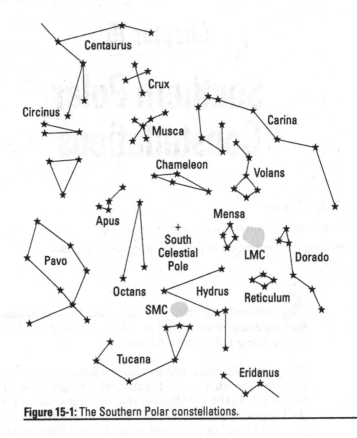

Figure 15-1: The Southern Polar constellations.

Crux

Crux, the Southern Cross (see Figure 15-2), is one of most recognisable patterns in the sky, even though it's the smallest of all 88 constellations.

The four stars that make up the Southern Cross are bright – 3 of them are in the top 30 brightest stars – so finding the Southern Cross is pretty easy. The Southern Cross lies next to the very bright stars *Rigel Kentaurus* and *Hadar* (α and β Centaurus).

Crux info

Abbreviation: Cru

Genitive: Crucis

Best month for observing in during the evening: May

Bright Stars List

Name	Bayer Designation	Type	Mag	Distance (ly)
Acrux	α Cru	B0IV	1.4	321
Mimosa	β Cru	B0III	1.25	352
Gacrux	γ Cru	M4III	1.59	88
	δ Cru	B2IV	2.79	364

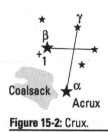

Figure 15-2: Crux.

The Southern Cross is one of the most prominent southern constellations, and so I list it here first, before the other more minor ones. Crux is one of the constellations you need to use to find due south. Draw a line down from the long axis of the Southern Cross and simultaneously draw a line that cuts *Rigil Kentaurus* and *Hadar* apart. Where those two lines meet is the southern pole (see Figure 2-3 in Chapter 2).

Look for the following in Crux. The numbers correspond to those in Figure 15-2:

1. The **Jewel Box cluster, NGC 4755,** is a beautiful sight through binoculars or a telescope. To the naked eye, the Jewel Box cluster appears as a dense smudge in the Milky Way, but magnify it and you'll see a field of brilliant blue-white stars, with a red star in the middle.

2. The **Coalsack** isn't really a faint fuzzy; rather, it's a dark patch on the Milky Way, looking as if someone's taken some stars away. In fact, the Coalsack is a cloud of dust that's obscuring the stars behind it. The Coalsack is very obvious to the naked eye and easy to spot given that it's right next to α and β Crucis.

Apus

Apus the Bird of Paradise (see Figure 15-3) is small and undistinguished, lying south of Triangulum Australe. Because Apus doesn't contain any very bright stars, it may be quite tricky to locate.

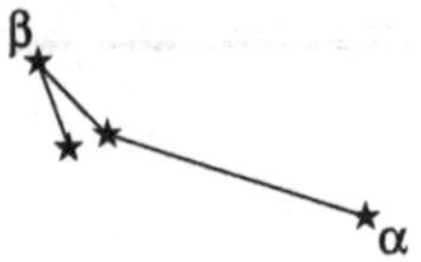

Figure 15-3: Apus.

Apus has no decent faint fuzzies.

Apus info

Abbreviation: Aps

Genitive: Apodis

Best month for observing in during the evening: July

Bright Stars List

Name	Bayer Designation	Type	Mag	Distance (ly)
	α Aps	K5III	3.83	411
	γ Aps	K0IV	3.86	159

Chamaeleon

Chamaeleon (see Figure 15-4) is another small, dim constellation. The Southern Cross's long axis points towards Chamaeleon, which makes it a bit easier to find.

Figure 15-4: Chamaeleon.

Chamaeleon has no decent faint fuzzies.

Chamaeleon info

Abbreviation: Cha

Genitive: Chamaeleontis

Best month for observing in during the evening: April

Chamaeleon has no stars brighter than magnitude 4.

Circinus

Circinus the Compasses (see Figure 15-5) is tiny and dim, but luckily it lies right beside *Rigel Kentaurus,* α Centaurus, the fourth brightest star in the sky, which makes finding it much easier.

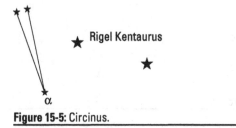

Rigel Kentaurus

α

Figure 15-5: Circinus.

Even though Circinus lies in the Milky Way, it has no faint fuzzies to look for.

Circinus info

Abbreviation: Cir

Genitive: Circini

Best month for observing in during the evening: June

Bright Stars List

Name	Bayer Designation	Type	Mag	Distance (ly)
	α Cir	F1V	3.18	2.11

Dorado

Dorado the Swordfish (see Figure 15-6) lies between the bright stars *Canopus,* α Carinae, and *Achernar,* α Eridani. Dorado isn't an especially recognisable constellation, but it's significant because it contains most of the Large Magellanic Cloud (LMC), looking like it's leaping out of the misty patch in the sky.

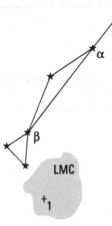

Figure 15-6: Dorado.

Look for these things in Dorado:

1. The **Tarantula Nebula, NGC 2070,** lies in the LMC, and is bright, visible to the naked eye, and looks great through binoculars and telescopes.

2. The **LMC** is the largest galaxy visible in the night sky (excluding the Milky Way). In fact, the LMC looks like a bit of the Milky Way that's broken off and is floating through space, but instead it's a galaxy in its own right. It's called the Large Magellanic Cloud after Ferdinand Magellan, the European explorer who first reported seeing it, in 1519. The LMC is actually only a dwarf galaxy, much smaller than the Andromeda Galaxy, M31 (it has only 1/100th the mass of the Andromeda Galaxy) but it's much nearer to Earth than the more distant M31 (only 160,000 light years compared with Andromeda's 2.5 million light years) and so looms bigger in the sky. Scanning the LMC with binoculars or a telescope is worth it because it's full of star clusters and clumpy patches, the Tarantula Nebula being the best one.

Dorado info

Abbreviation: Dor

Genitive: Doradus

Best month for observing in during the evening: January

Bright Stars List

Name	Bayer Designation	Type	Mag	Distance (ly)
	α Dor	A0III	3.30	176
	β Dor	F4Ia	3.76	1038

Hydrus

Hydrus the Little Water Snake (see Figure 15-7) is much smaller than the giant Hydra, the largest of all the constellations. Hydrus lies south of the bright star *Achernar,* α Eridani, and near the Small Magellanic Cloud (SMC), so finding the general area of Hydrus isn't hard; joining the dots into anything resembling a snake is much harder!

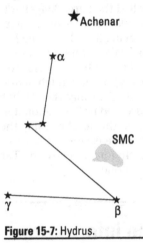

Figure 15-7: Hydrus.

Hydrus contains no good faint fuzzes. The SMC is nearby, but actually lies in the neighbouring constellation Tucana.

Hydrus info

Abbreviation: Hyi

Genitive: Hydri

Best month for observing in during the evening: November

Bright Stars List

Name	Bayer Designation	Type	Mag	Distance (ly)
	α Hyi	F0V	2.86	71
	β Hyi	G2IV	2.82	24
	γ Hyi	M2III	3.26	214

Mensa

Mensa the Table Mountain (see Figure 15-8) lies south of the Large Magellanic Cloud (LMC); in fact, the LMC spills into this constellation, which at least makes it a bit easier to spot. Mensa has no bright stars in it, though.

LMC

Figure 15-8: Mensa.

You'll not find a lot to see in Mensa, except for the LMC. (See 'Dorado' for a fuller description.)

Mensa info

Abbreviation: Men

Genitive: Mensae

Best month for observing in during the evening: January

Mensa has no stars brighter than magnitude 4.

Musca

Musca the Fly (see Figure 15-9) is a small constellation lying to the south of the Southern Cross. The stars of the long axis of Crux point towards Musca, making it easier to find.

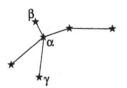

Figure 15-9: Musca.

Musca contains no significant faint fuzzies.

Musca info

Abbreviation: Mus

Genitive: Muscae

Best month for observing in during the evening: May

Bright Stars List

Name	Bayer Designation	Type	Mag	Distance (ly)
	α Mus	B2IV	2.69	306
	β Mus	B2V	3.04	311
	γ Mus	B5V	3.84	324

Norma

Norma the Set Square (see Figure 15-10) is yet another dim, unimpressive southern polar constellation. Norma lies south of the tail of Scorpius.

Figure 15-10: Norma.

Once again, you won't see much in this constellation.

Norma info

Abbreviation: Nor

Genitive: Normae

Best month for observing in during the evening: June

Norma has no stars brighter than magnitude 4.

Octans

Octans the Octant (see Figure 15-11) is rather dim but is distinguished by being the constellation that houses the south celestial pole. You won't see a bright South Pole star like you can in the north, so finding south is a bit trickier. However, you can use Crux and Centaurus to find the approximate area and therefore to find Octans, too.

Octans info

Abbreviation: Oct

Genitive: Octantis

Best month for observing in during the evening: October

Bright Stars List

Name	Bayer Designation	Type	Mag	Distance (ly)
	ν Oct	K0III	3.73	69

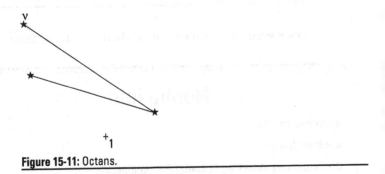

Figure 15-11: Octans.

The South Celestial Pole, numbered 1 in Figure 15-11, lies in the constellation Octans, near the faint star σ Octantis, which, at magnitude 5.6, is at the limit of what most people can see with the naked eye under dark skies. As a result, it is too faint to be marked on this star chart.

Pavo

Pavo the Peacock (see Figure 15-12) contains one bright star, also called *Peacock,* α Pavonis, which lies half way between Grus and Ara. The rest of Pavo is quite hard to spot, but a rectangular-looking tail completes the peacock shape.

Pavo info

Abbreviation: Pav

Genitive: Pavonis

Best month for observing in during the evening: August

Bright Stars List

Name	Bayer Designation	Type	Mag	Distance (ly)
Peacock	α Pav	B2IV	1.94	183
	β Pav	A5IV	3.42	137

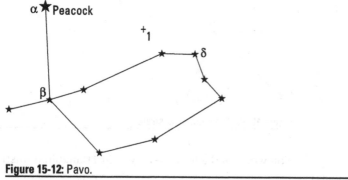

Figure 15-12: Pavo.

Look for **NGC 6752** (numbered 1 in Figure 15-12), a globular cluster just visible to the naked eye in good conditions. You can find NGC 6752 on a line between ζ Pavonis and α Pavonis (*Peacock*).

Triangulum Australe

Triangulum Australe (see Figure 15-13) is the southern counterpart to the northern constellation Triangulum, but it's a bit easier to find because its main stars are brighter. Triangulum Australe lies near *Rigel Kentaurus* and *Hadar* (α and β Centaurus).

Triangulum Australe info

Abbreviation: TrA

Genitive: Trianguli Australis

Best month for observing in during the evening: July

Bright Stars List

Name	Bayer Designation	Type	Mag	Distance (ly)
Atria	α TrA	K2IIb	1.91	415
	β TrA	F2III	2.83	40
	γ TrA	A1V	2.87	183

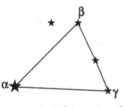

Figure 15-13: Triangulum Australe.

You won't find a lot to keep you in Triangulum Australe.

Tucana

Tucana the Toucan (see Figure 15-14) lies near the bright star *Achernar*, α Eridani, and the Small Magellanic Cloud (SMC). Tucana isn't very distinct, but *Achernar* and the SMC may help you find it.

Look for these things in Tucana. The numbers in the list correspond to the numbers in Figure 15-14:

1. **47 Tucanae** is widely recognised as the second best globular cluster in the sky after ω Centauri, NGC 5139. 47 Tucanae is right beside the SMC and is visible to the naked eye. If you use a medium telescope, you should be able to make out individual stars in 47 Tucanae.

2. The **SMC,** like its bigger brother, the Large Magellanic Cloud (LMC) (see 'Dorado' above), looks like part of the Milky Way that's broken off, but in fact is a dwarf galaxy near the Milky Way. Like the LMC, the SMC is a dwarf galaxy around 200,000 light years away (a little farther than the LMC) and has a slightly smaller mass than the LMC, so contains fewer stars (only a few billion).

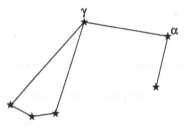

SMC

Figure 15-14: Tucana.

Tucana info

Abbreviation: Tuc

Genitive: Tucanae

Best month for observing in during the evening: November

Bright Stars List

Name	Bayer Designation	Type	Mag	Distance (ly)
	α Tuc	K3III	2.87	199
	γ Tuc	F1III	3.99	72

Volans

Volans the Flying Fish (see Figure 15-15) lies south of Carina the Keel, near the bright star *Miaplacidus*, β Carinae.

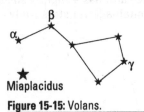

Miaplacidus

Figure 15-15: Volans.

You won't see any decent faint fuzzies in Volans.

Volans info

Abbreviation: Vol

Genitive: Volantis

Best month for observing in during the evening: February

Bright Stars List

Name	Bayer Designation	Type	Mag	Distance (ly)
	α Vol	Am	4.00	124
	β Vol	K2III	3.77	108
	γ Vol	G8III	3.78	142

Part IV
The Part of Tens

The 5th Wave   By Rich Tennant

SUDDENLY, OUT OF THE AUTUMN SKY CAME THE CONSTELLATION KNOWN AS RHONDA

In this part...

Stargazers have so much choice! The sky is teeming with interesting things to look at, so where do you start? Never fear, after reading the Part of Tens you'll have two lists of objects to go out and find; a Top-Ten list of great astronomical objects to get you started, and then a Top-Ten list of dark sky objects, things you'll only see once you leave the bright city lights behind you and head somewhere dark!

Chapter 16

Ten Targets for New Stargazers

In This Chapter

▶ Picking your first stargazing targets

▶ Finding Jupiter's moons and Saturn's rings

▶ Identifying your brightest faint fuzzies

As an excited new stargazer, you'll want to get outside and start straight away. This chapter contains a handy list of targets that you can tick off on your way to mastering the night sky.

The Moon

Earth's closest celestial neighbour is a splendid sight in the night sky, even if it does spoil the view of fainter objects. If you're hunting dark-sky targets (see Chapter 17), you'll need to wait for a night when the Moon isn't in the sky, but while you wait, you should take advantage of the great display the Moon puts on. Every night is different as the Moon waxes and wanes and the terminator line between light and dark moves over its face and different craters and mountains become visible. Even small binoculars give a great view of the Moon, but through a telescope nothing else can beat the Moon.

The International Space Station

The giant International Space Station (ISS for short) whizzes around the Earth every 92 minutes, and on some passes it catches the Sun's light at just the right angle to light it up, letting you see it from Earth. These passes happen like clockwork, so if you know where and when to look, you can always spot the ISS. Find out when the ISS passes over on www.heavens-above.com.

Saturn's Rings

Saturn's beautiful ring system shines like a jewel through even small telescopes. Something is magical about seeing these rings for the first time *live* rather than on a page or a screen. Admittedly, the view through your telescope – no matter how big it is – won't look as visually stunning as the pictures sent back by the Cassini space probe, but they're a *real* view, at the end of your telescope, of this beautiful planet hanging in space.

Jupiter's Moons

It's a toss-up which is the best planet to see through a telescope, but for me Jupiter edges out Saturn. Sure, Jupiter doesn't have amazing rings, but it does have cloud bands, a Great Red Spot, and four giant moons orbiting it. If you watching Jupiter through a telescope over the course of a night or on several nights over a week, you'll see these objects change position as the king of the planets spins. When you see Jupiter's moons for the first time, you see things that changed the world forever, because they proved to Galileo that not everything in the cosmos orbited the Earth, contrary to what was generally believed at the time.

'Canals' on Mars

Mars is a staple favourite of sci-fi fans, and it hangs orange-red in a dark night sky. However, you need a medium or large telescope to see much on the surface of the Red Planet, but making the effort is worth it. Use your imagination to trace dark stripes and patches on the surface of the planet and imagine how those areas may have looked to the first astronomers who saw them – like the markings left behind by an intelligent civilisation. Now astronomers know they're natural colourations of rocks on Mars, but that fact hasn't stopped over a century of speculation about Martians and a fascination with the Red Planet, which may be the next stop for the human race in its exploration of the solar system.

Phases on Venus

Venus is the brightest planet – and the most beautiful when seen hanging in a deep twilight sky after sunset or before sunrise. Look through a large pair of binoculars or a telescope at Venus at its brightest, and you'll see it has a crescent shape, looking like a tiny version of the Moon. You can track Venus's other phases as it orbits the Sun.

 You should always make sure that the Sun is below the horizon when scanning that part of the sky with binoculars or a telescope.

Elusive Mercury

Seeing Mercury at all is tricky, as it's seldom visible in the glare of the Sun. If you can catch sight of Mercury, you'll see it shining like a bright-ish star in twilight, but you'll be in a select club of people ever to have seen it. You may see phases of Mercury like you see for Venus, but you'll need a telescope to see them, and you'll need to take extra-special care that the Sun is well and truly below the horizon if you value your eyesight.

Sunspots

After you buy your first telescope, you should supplement it with a solar filter, a cover for its front end. (Never use solar filters that go on the eyepiece.) This filter lets you look safely at the giant ball of gas in the daytime sky.

On the surface of the Sun are tiny black specks known as *sunspots,* regions of cooler gas. Depending on whether the Sun is very active, you'll see different numbers of sunspots. Sometimes the surface can be peppered with lots and lots of little dark spots, each one larger than the Earth.

The Big Dipper, the Southern Cross

Stargazers in the northern hemisphere quickly come to know and love the Big Dipper, and their southern cousins feel the same about the Southern Cross, both of which you can use to find your pole, north or south.

The first thing that any good stargazers look for when heading outside at night is their polar signpost: find north or south, and use this guide to find everything else!

The Orion Nebula

You can't find this stunning faint fuzzy in the sky all year round, but when the Orion Nebula is above the horizon, you'll quickly come back to it time and time again.

Orion itself is one of the most distinct of all the constellations, and so finding the Orion Nebula, M42, is easy:

1. **Find Orion's stars *Betelgeuse* and *Rigel*, two of the brightest stars in the sky.**

2. **Find Orion's Belt half way between *Betelgeuse* and *Rigel*.**

3. **Find Orion's sword hanging from his belt.**

   The Orion Nebula is the faint fuzzy in the middle of the sword.

The Orion Nebula is visible to the naked eye, even in sites with a bit of light pollution, as a fuzzy patch in Orion's sword. Through a telescope, you begin to see detailed structure among the gases in this giant cloud of star-forming hydrogen, as well as the stars that light it up from the inside.

Chapter 17

Ten Things to Look for under a Dark Sky

*I*n this chapter, I describe ten things to look for the next time you're lucky enough to be out stargazing under a dark sky. Most people live under a perpetual shroud of light pollution from towns and cities, so you may have to travel to see everything in this chapter. In fact, you'll certainly have to travel to see them all, because some of them are only visible from certain parts of the Earth and at certain times of year. Chase down those dark skies; you won't be disappointed.

The Number of Stars

The first thing you'll notice as you move from city to country-side is the sheer number of stars overhead on a clear night. You should make a habit of counting how many stars you can see in a familiar constellation such as Orion, from different locations, and that number will help you work out just how good (or bad) your sky is.

In a city, you may see only a hundred or so of the brightest stars (those brighter than magnitude 4, say), while under darker conditions, you may see more than a thousand.

Once you're out somewhere really dark, the only limit is your eyesight. Many thousands of stars are up there if you can catch them.

The Milky Way from Horizon to Horizon

The Earth's galaxy, the Milky Way, is an impressive sight under a dark sky. You can catch glimpses of the Milky Way from the outskirts of towns and cities, from where it may appear as a faint patch overhead. Get away from light pollution, though, and you can see the band of the galaxy stretching across the sky.

If you ever get somewhere *really* dark, then you'll begin to see structure in the Milky Way – dark bands and blobs that come from clouds of dust in the galaxy blocking out the starlight of the stars behind. One of the most breathtaking sights in the night sky is the Milky Way sitting overhead in an arch stretching from horizon to horizon.

The Andromeda Galaxy, M31

The next closest big galaxy to Earth's is the Andromeda Galaxy or, to give it its catalogue number, M31. If you know where to look, you can catch glimpses of the Andromeda Galaxy from light-polluted skies, but you can see it best when it's darker overhead.

The Andromeda Galaxy should look like a cigar-shaped smudge to your naked eyes. Through a good pair of binoculars or a telescope you can see more detail; perhaps you'll catch a glimpse of the Andromeda Galaxy's nearby companion galaxies, M31 and M110, or even see dark dust lanes within the main galaxy itself.

Remember that what you're looking at is a galaxy that's even bigger than the Milky Way, containing perhaps *a trillion* (1,000,000,000,000) stars. Aliens living on a planet orbiting a star in the Andromeda Galaxy would look up into their night sky and see a faint Andromeda-like smudge, too – our Milky Way galaxy seen from 2.5 million light years away.

The Triangulum Galaxy, M33

If Andromeda is an easy galaxy to find, its neighbour, the Triangulum Galaxy, or M33, is much more elusive. The Triangulum Galaxy is much fainter and therefore harder to find if any light pollution is around, but that makes it a great benchmark for dark skies. Try to chase down the Triangulum Galaxy to see whether your skies fall under the category of really dark.

You'll have to travel out of the glow of suburbia to see M33, and even then it's hard to spot. Even under clear rural skies you'll have to use averted vision to catch it at all, but if you want to see the Triangulum Galaxy when you're looking straight at it, then you have to head to a truly dark-sky site, many miles from the nearest town.

M33 looks like an even fainter version of the Andromeda Galaxy, almost impossibly faint, because it's smaller, farther away, and has less than a tenth the number of stars than Earth's nearby giant galaxy.

The Seven Sisters

A great test of your eyesight – and your dark skies – is the faint fuzzy known as the Pleiades, or the Seven Sisters (M45 to its friends). Under city skies, you see it looking like a distinct fuzzy patch. You may be able to pick out the brighter stars, but you need to be somewhere dark; you'll have to wait for a long time to let your eyes adapt before you'll see all seven.

Some people with excellent eyesight claim to be able to see more than seven stars in the Pleiades. Have a go, and after you've counted the stars with your naked eye, see how many you can spot through a pair of binoculars. Wow, right?

Aurorae

The elusive, shimmering colours of the northern or southern lights, the *aurora borealis* or *aurora australis*, are chased away by light pollution, and so you need to head out of town to see them. But your journey may not end there. In order to see the aurora best, you have to head to high latitudes.

In the northern hemisphere, you can see the aurorae best from the Arctic, northern Europe, Scandinavia, northern Russia, Canada and northern America states; in the southern hemisphere, you see them best from Antarctica, southern South America, New Zealand and southern Australia.

The aurorae don't happen every night, only after strong solar storms, so keep an eye on that space weather!

Meteor Showers

Not strictly a dark-sky phenomenon, meteors are visible even in cities, but you can see lots more of them the darker your sky is. On any night, you can expect to see a few shooting stars under dark skies, but during meteor showers this rate increases dramatically. At the peak of an active meteor shower like the Perseids or the Geminids, you can expect the maximum rate to get up to 100 per hour – that's 100 shooting stars an hour, more than one per minute! But light pollution will drown out the fainter ones, so in suburbia you may see only 30 per hour, and in a city centre perhaps only 10 an hour.

Zodiacal Light

If you're lucky enough to be stargazing from a good dark site in spring or autumn, you may catch a glimpse of the elusive *zodiacal light*. Zodiacal light is light from the Sun scattering off particles of dust in the solar system and reflecting back to your eyes.

You can see zodiacal light only just after sunset twilight or just before sunrise twilight, and your window of opportunity is very narrow. Time your zodiacal light hunting for the end of evening astronomical twilight in spring or the beginning of morning astronomical twilight in autumn.

The zodiacal light looks light a faint cone of light – much less bright than the Milky Way – that stretches along the line of the zodiac, or ecliptic. Under exceptional dark skies, you can see a yellowish colour to the zodiacal light, which can stretch across the whole sky in an arc and can even get bright enough to cast a shadow.

Gegenschein

The elusive partner to the zodiacal light is called *Gegenschein*. The name is German, meaning counter-shine, and it's light from the Sun reflecting off particles of dust that lie directly opposite the Earth from the Sun, in a part of the sky called the *antisolar point*. Because you see these dust particles lit face-on, the Gegenschein can be a bit brighter than the zodiacal light elsewhere in the sky.

You need to be somewhere *really* dark to see Gegenschein, and you need to look in the right direction. Follow the cone of the zodiacal light up, along the ecliptic, and look for a faint oval glow; this is *Gegenschein,* and not many people get to see it.

Airglow

If you're somewhere with no manmade light pollution in the sky, you may notice that the sky *still* isn't pitch black. That's partly due to something called *airglow,* which is the emission of light by the molecules of gas that make up the atmosphere. Airglow is a perfectly natural phenomenon, but it means that the sky never gets truly dark. You may catch a glimpse of airglow low on the horizon, where you're looking through more atmosphere, around one fist-height or 10 degrees above the horizon. It'll look like a faint blue glow in the air.

Index

• A •

About the Author

Steve Owens, M.Sci. (Hon), an astronomer and professional science communicator with two decades of experience, was the 2010 recipient of the Federation of Astronomical Societies' Erik Zucker Award for 'outstanding work for the astronomical community', the 2010 recipient of the Campaign for Dark Skies Award for 'meritorious efforts in preserving dark skies', and the 2011 recipient of the International Dark-Sky Association's Defender Award. He was planetarium manager at Glasgow Science Centre from 2004 to 2008, and the UK manager for the International Year of Astronomy 2009, during which time he helped establish Europe's first International Dark Sky Park in Galloway, Scotland. Since then he has helped set up many more such places in the UK, and around the world. He is the current chair of the International Dark-Sky Association's International Dark Sky Places Development Committee, which helps promote and develop such dark sky places. He is a dark skies consultant, a freelance stargazer, and astronomy writer.

In the course of his career as an astronomer and stargazer he has led stargazing dinners in the sand dunes of the Namib Desert, given astronomy lectures on the remote island of Saint Helena, run stargazing night classes in the dark depths of the highlands of Scotland, and delivered hundreds of planetarium shows to audiences of all ages. He has written many articles on astronomy for UK newspapers and magazines, blogs regularly at darkskydiary.co.uk, and appears on Twitter as @darkskyman.

He is a graduate of the University of Glasgow, where he received a first class honours degree in astronomy. He still lives in Glasgow, with his partner Samantha and their son Elliot.

Dedication

To Samantha and Elliot, with love.

Author's Acknowledgments

I'd like to thank my family and friends who supported me while I was writing this book, and to the staff as Wiley Publishing who provided expert guidance in crafting it into its final form.

I'm indebted to the astronomers at the University of Glasgow – in particular Professor Martin Henry, Dr David Clarke, and Professor John Brown – who many years ago helped me take my first steps in my career as a stargazer, to Professor Ian Robson of the UK Astronomy Technology Centre, and also to the staff of Glasgow Science Centre who housed me during much of the writing of this book, and in whose wonderful planetarium I wiled away many a cloudy night. I'd also like to thank Professor Carolin Crawford for the insights and suggestions that made this book all the more accurate.

Thanks also to Nick Howes, who provided images which some of the drawings in this book are based on. I drew the star maps in this book with the help of many great star guides and atlases, in particular Roger Sinnott's excellent *Pocket Sky Atlas* and Will Tirion's superlative *Sky Atlas 2000.0.*

Publisher's Acknowledgments

We're proud of this book; please send us your comments at http://dummies.custhelp.com. For other comments, please contact our Customer Care Department within the U.S. at 877-762-2974, outside the U.S. at (001) 317-572-3993, or fax 317-572-4002.

Some of the people who helped bring this book to market include the following:

Project Editor: Simon Bell

Commissioning Editor: Kerry Laundon

Assistant Editor: Ben Kemble

Development and Copy Editor:
Kelly Ewing

Technical Editor:
Professor Carolin Crawford

Proofreader: Mary White

Production Manager: Daniel Mersey

Publisher: Miles Kendall

Cover Photos: © Peter Burnett / iStock;
© Science Photo Library / Alamy

Cartoons: Rich Tennant,
www.the5thwave.com

Project Coordinator: Kristie Rees

Layout and Graphics: Brent Savage

Proofreaders: Lauren Mandelbaum,
Susan Moritz

Indexer: Potomac Indexing, LLC